David Brown

The Apocalypse

Its structure and primary predictions

David Brown

The Apocalypse
Its structure and primary predictions

ISBN/EAN: 9783744747493

Printed in Europe, USA, Canada, Australia, Japan

Cover: Foto ©berggeist007 / pixelio.de

More available books at **www.hansebooks.com**

THE APOCALYPSE:

ITS STRUCTURE AND PRIMARY PREDICTIONS.

BY

DAVID BROWN, D.D.,

Principal of the Free Church College, Aberdeen.

London:
HODDER AND STOUGHTON,
27, PATERNOSTER ROW.

MDCCCXCI.

PREFACE.

IT is with more diffidence than I can expect to get credit for that I issue this booklet. I have not written a line on the subject of Prophecy since, nearly fifty years ago, I published a book on the Second Advent. My studies since then have lain in a more extended direction. But the successive editions of that book, and the increased attention given to Prophecy, in its bearing on the *time* when the Second Coming of our Lord is to be expected, led to a desire, repeatedly expressed, that I should write something on the Book of Revelation, and, from the other side of the Atlantic, that I should write a commentary on that Book—a thing which I need hardly say I did not listen to for a moment, nor would

I consent to take up my pen on the subject at all. But after repeated solicitations I at length committed myself so far as to write to the able Editor of the *Expositor* that I might try an A B C of the Apocalypse—perhaps promised to do it. Whereupon it was announced as among the subjects on which papers might be expected in the forthcoming years. Yet only after two years did I send a tentative article on the *Date* of the Apocalypse, which was followed up by one on the *Design* of the Apocalypse; after which I got so into the subject that the issue is the present volume.

This is the whole history of the book, and I mention it only that the reader may understand what could have induced one at my age to close my literary work with this subject. On the main lines of the Book of Revelation my mind had long been made up, and is now as decided as ever. As a contribution, therefore, to the understanding of the mind of the Spirit in it, I thought it might be a duty not to refuse to give others the benefit of my own studies in it, such as they were.

I will only add that oversights may probably be found here and there; but when one is nearly eighty-eight years of age, while his feeble eyesight can receive no aid from artificial light, such things will be pardoned. In fact, should none appear, it is due to the intelligent pains taken upon the proofs in the printing office, which I must here gratefully acknowledge.

March 4th, 1891.

CONTENTS.

INTRODUCTION:

	PAGE
(1) Authenticity of the Apocalypse	3
(2) Date of the Apocalypse	7
(3) Design of the Apocalypse	26
Addendum: Sir W. Hamilton's Attack on the Apocalypse	35
STRUCTURE OF THE APOCALYPSE	67
VISION OF THE SEVEN SEALS	70
PRINCIPLE OF INTERPRETATION	75
FIRST SEAL	79
FIFTH SEAL	79
SIXTH SEAL	81
THE WOMAN AND THE DRAGON	83
THE WAR IN HEAVEN	86
PREPARATIONS FOR THE OPENING OF THE SEVENTH SEAL	91
THE SEVENTH SEAL AND SEVEN TRUMPETS	96

CONTENTS.

	PAGE
THE DRAGON'S NEW POLICY	99
(1) "Woe for the earth"	100
(2) "Woe for the sea"—Rise of the "beast"	102
CHARACTERISTICS OF THIS BESTIAL POWER	104
MEASURING THE TEMPLE, THE ALTAR, AND THE WORSHIPPERS	112
THE TWO WITNESSES	115
THEIR MARTYRDOM, RESURRECTION, AND ASCENSION	118
THE SEVEN VIALS, WITH THE CHORAL HYMNS	126
THE KEY TO THE MYSTERY	136
SUMMARY	145
CONFIRMATORY PREDICTIONS:—	
2 Thess. ii. 1-12; 1 Tim. iv. 1-5	149
THE FALL OF BABYLON AND SUMMONS TO COME OUT OF IT ERE IT FALL	161
DIRGE OVER THE FALL OF BABYLON	165
HALLELUJAHS OVER ITS FALL	166
THE LAST WAR, AND END OF CHRIST'S PUBLIC ENEMIES	170
THE THOUSAND YEARS	176
THE THOUSAND YEARS' REIGN WITH CHRIST	180
THE REST OF THE DEAD	191

	PAGE
SATAN'S LAST EFFORT AND FINAL DEFEAT	195
THE GENERAL JUDGMENT	198
THE LAST THINGS	200
CONCLUDING REMARKS	202
ADDENDA	211

1

INTRODUCTION.

I.

AUTHENTICITY OF THE APOCALYPSE.

For detailed evidence of this I must refer to the Introductions to the New Testament, and especially to this book, which of late years have been abundant. But a few *notes* of my own may not be out of place here.

1. It is a remarkable fact that, of the six books of the New Testament whose authenticity was doubted in the early Church, the doubts about five of them—the Epistle of James, the Second of Peter, the Second and Third of John, and that of Jude—existed from the first, and only gave way in the fourth century; whereas the authenticity of the Apocalypse only began to be doubted in the third century, when Dionysius, bishop of Alexandria, went down to Egypt to refute the Chiliasts—the millennarian followers of Nepos, the warm-hearted bishop of Arsinoé. And it was only when hard pressed by their Apocalyptic arguments, that he had to confess that he had great doubts about the authenticity of that book. Not that he could name any one who had done so before him,—but because of its obscurity; the grossness of its millennarianism; the barbarisms or solecisms of its Greek (so different in every respect from the apostle John's); the writer's naming himself "John," whereas the apostle seems studiously to avoid naming himself; and the uncertainty as to what John it

was who wrote it—whether it may not have been some Presbyter of that name.*

It says not a little, I think, for the critical acuteness of this bishop, that he has anticipated all that up to this time has ever been urged against the authenticity of this book. The objections themselves are all internal, with one exception; and they will be sufficiently met under the next head —the *Date* of the Apocalypse. The one objection not there touched on is, that the writer frequently names himself "John," which the apostle John in his acknowledged writings never does. But the answer to that is easy. Those who read the Fourth Gospel and the three Epistles have no need to be told that the writer expects to be recognised as the beloved disciple; whereas the Apocalypse is throughout a record of what no one was, or could be, an eye and ear witness to but the writer himself. He was therefore bound to tell his readers who he was, and how and in what circumstances these strange revelations were made to him—and this is just what he tells us.

As for Dionysius, he was one of the "Allegorists," who were driven to interpret the book in a spiritual sense to meet the Literalists. His master was *Origen*, the greatest Biblical scholar which the Church produced. In his travels he made it his business to collect all the information he could on Biblical subjects—as witness his great *Hexapla* of the Old Testament. *His* testimony, therefore, on the authenticity of the Apocalypse is of peculiar value. His works (3 vols., folio), I have carefully examined, and have found as many as fifty, either quotations from or references to the Apocalypse as a genuine work of the apostle John, without a hint of any doubt on the subject having existed.

---

* *Euseb.,* H. E., vii. 24, 25.

2. *Papias*, bishop of Hierapolis (in the vicinity of the seven churches), was within the circle of the apostle John's immediate disciples; and no one who reads what Eusebius tells us of his stories about what would be seen during the thousand years, can doubt that they were all spun out of what he supposed the Apocalypse to predict about that period.

3. *Justin Martyr* (A.D. 140—160) explicitly ascribes the Apocalypse to the apostle John; and in his dialogue with Trypho the Jew he expresses his own expectations of what the earth would be during the thousand years.

4. *Melito*, bishop of Sardis (one of the seven churches) and a contemporary of Justin Martyr, wrote several works, which Eusebius enumerates, one of which (or two) is on the devil and the Revelation of John.

5. The priceless *Letter of the Churches of Lyons and Vienne* to the brethren of the Asiatic Churches, giving them, at their request, a thrilling account of the sufferings of the Christians of Gaul, and their heroic endurance of martyrdom, in the persecution set on foot by Marcus Antoninus (A.D. 177). That letter—preserved almost entire by *Eusebius* (H. E., v. 1, 2)—tells us how the martyrs "followed the Lamb whithersoever He would go"; and when we find it quoting the words, "He that is unjust let him be unjust still, and he that is holy let him be holy still," we can be at no loss to know what book had endeared itself to those suffering Christians.

But if the authenticity of the Apocalypse was recognised so early, how was it not in either the *Old Syriac* or the *Old Latin* versions—now known to have been made not later than the middle of the second century? That question is easily answered. The books most read in the public services, and most prized, would of course be the first to be translated—the four Gospels; and then gradually all the rest. But, as we may be quite sure that the Apocalypse would, no more than now, be read in public (and for the same obvious reason), we may be sure that it would remain untranslated for some time. For the same reason, fewer copies of the original would be made than of all the other books, and for this reason too, of all the books of the New Testament now extant, copies of the Apocalypse are far the fewest.

But unread in the public services though it naturally was, it was far from being little studied and valued in private. The bitter persecutions of the Christians by successive emperors no doubt raised the cry to heaven, *Quousque Domine!* ("But Thou, O Lord, how long?"), and made this book of the New Testament so tenderly encouraging to the faith and patience of the saints, its beautiful promises "to him that overcometh," its glowing assurances of final triumph, and its bright pictures of the reign of Christ on the earth. And what though an excessive literalism in picturing the millennial future took hold, as it certainly did, of its ardent students, who can wonder? And as little need we wonder at their opponents being driven into the opposite extreme.

II.

DATE OF THE APOCALYPSE.*

AMONG competent judges the difference on this subject lies between two periods : *the reign of Nero,* and shortly before the destruction of Jerusalem, about A.D. 68 ; or *the reign of Domitian,* and shortly before his death, about A.D. 95 or 96.

Of *external evidence* for the former date there is absolutely none. In fact, this date was never heard of till the sixth century, and even then only in the superscription to a Syriac version of the book supposed to be of that date. After that we hear nothing of it till, in the twelfth century, we find Theophylact assigning it to the reign of Nero.

But what says ecclesiastical history to the later date? The great witness, as he is the primary one, is *Irenæus,* bishop of Lyons A.D. 177 to *circa* 202. To Gaul he came from Asia Minor, where he tells us he was a hearer of Polycarp, bishop of Smyrna and disciple of the apostle John. In his great work *Against Heresies,* we find him discussing the two readings of "the number of the beast" (Rev. xiii. 18), whether in the original text it was 666 or 616. He says that in all the approved and ancient copies (ἐν πᾶσι ταῖς σπουδαίαις καὶ ἀρχαίαις ἀντιγράφαις) the reading was 666, and that this reading was attested by those who

---

* From the *Expositor* (1889), vol. x. (3rd series), pp. 272-88.

had seen John face to face (καὶ μαρτυρούντων αὐτῶν ἐκείνων τῶν κατ' ὄψιν τὸν Ἰωάννην ἑωρακάτων). The importance attached to this reading lay in the belief that this number enigmatically pointed to the expected antichrist, whose name (he says) he will not speak of confidently, "because had it been necessary to name him at the present time, it would have been declared by him who saw the Revelation; *nor was it long since it had been seen, but almost in our own generation* (οὐδὲ γὰρ πολλοῦ χρόνου ἑωράθη ἀλλὰ σχεδὸν ἐπὶ τῆς ἡμετέρας γενεᾶς), *about the end of Domitian's reign.*"

This very important statement is twice quoted *verbatim* by *Eusebius* (H. E., iii. 18 and v. 8), and the value of it is so felt by the advocates of the early date, that they make every effort to break it down; while all subsequent testimony is regarded as but an echo of this one, and therefore of no value. We must weigh it, then, and all the more because the *date* of the book has an important bearing on the interpretation of it.

Observe, then, that Irenæus "saw and heard" Polycarp in his youth, or early manhood (ἐν τῇ πρώτῃ ἡμῶν ἡλικίᾳ, iii. 4); and he so describes him as to show what a deep impression that venerable Father had made upon him—an impression of his person as well as his teaching—as may be gathered from a remarkable passage in his "Letter to Florinus." He is there reasoning against certain heresies, and he appeals to the testimony of Polycarp, whose disciples Florinus and he had been. "For I saw thee while I was yet a youth (παῖς ὢν ἔτι) in Asia Minor with Polycarp. For impressions made in youth are better remembered than those made quite recently. For what we have been in our youth grows with our spirit, and gets incorporated with it, insomuch that I could even tell the place where the blessed Polycarp sat when discoursing, his exits and entrances, his manner of life and the appearance of his person, his addresses to the people, and his familiarity with John and others who had seen the Lord, which he related to us, and their sayings which he reported."*

May I not appeal to those who will candidly weigh these statements, whether they do not show that Irenæus was speaking *from knowledge* of the fact, when he says that the Revelation was seen not long since, but almost in his own generation, near the close of Domitian's reign?

Coming next to the *internal* evidence for the Neronic date—for it has nothing else to rest on—let us see what the book itself has to say to the question.

1. In the first, the introductory chapter, the seer tells us how, when in the rocky isle of Patmos, in the Ægean Sea, banished there "for the word of God and for the testimony of Jesus Christ," he was "in the spirit on *the Lord's day.*" Observe the testimony to the late date of this book which crops out, quite incidentally, at the very outset. Up to the date of the last of the Pauline Epistles, the only name for this day current among the Christians was "the first day of the week." Now, since (according to Jerome) the apostle Paul was beheaded in the fourteenth year of Nero's reign (A.D. 68), it must have been after that, and probably some years after, ere this most appropriate abbreviation came into such *established* use as is implied here. And if this is true, it disposes at once of the Neronic date.

2. The glaring difference between the Greek of the Apocalypse and that of the Fourth Gospel has led one class of critics to believe that both works cannot have come from the same author; while others (believing critics),

---

* *Irenæi Opp.*, ed. Stieren, 1883 (pp. 822, 823).

holding that both came from the pen of the apostle John, explain the peculiar style of Greek in which the Apocalypse is written by its early (Neronic) date, when the apostle was less familiar with the use of the language than when he wrote his Gospel. This was Dr. Westcott's view. But is this the only way for accounting for the *solecisms* of the Apocalypse? Startling they certainly are, both in their number and in their harshness; but that they are no proof of the writer's inability to write good Greek is freely admitted, and indeed is evident from his accuracy in other places. The only question, then, is, Must we explain it by his immaturity in the use of the Greek language? If so, you will have to explain how this immaturity does not show itself from beginning to end. And, what is harder still, you will have to show how so raw a hand, as you suppose the writer to be, was able to *coin* such compound words as ποταμοφόρητος (xii. 15), "river-borne" ("carried away by the stream," R.V.); and μεσουρανήμα (viii. 13, xiv. 6, xix. 17), "mid-heaven," found only in one medical writer of about the third century; and χαλκολίβανον (i. 15, ii. 18) "burnished brass" (R.V.). This kind of coinage seems to me to put an end to the theory of unfamiliarity with the use of the language, and, so far as that is concerned, to the necessity of an early date to the Apocalypse.

How else the solecisms are to be accounted for, I pretend not to explain. But I may be pardoned for throwing out this conjecture. Suppose the seer, being "in the Spirit," and writing under this inspiration, should find that in the rapid flow of his words these abnormal forms had dropped unsought from his pen, half dithrambically, but on observing this, had thought it right to leave them uncorrected, is there anything incredible or improbable in this? Be this, however, as it may, if it is not to be traced to ignorance of the language, it has no bearing on the date of the book.

But, it may be said, it is not on the solecisms of the book only that we rest; the whole *style* of Greek used here differs from that of the Fourth Gospel. True enough; but why? Not because of any difference of *date*, but because the *subject-matter* required a totally different style of writing. Every one knows the difference between prose and poetry. Poets studiously avoid ordinary, familiar forms of expression, and in the choice of words and phrases they go out of their way to find whatever is rare, startling, figurative. Now, the prophetic style, while it has all these characteristics of real poetry, has a boldness and intensity peculiar to itself. Dealing, as it does with the unseen, the celestial and infernal, the transporting and the terrifying, with what stirs the soul as earthly things cannot, it rises to heights and sinks to depths of its own. And if this is to be seen in all Hebrew prophecy, in the Apocalypse it stands out unrivalled.

As to the avoidance of familiar words and phrases, let any one, with his Greek Testament in hand, observe the number of uncommon words and phrases, *evidently selected as such*, in the Apocalypse, and he will be convinced, I think, that not any difference of date will explain this, but that it is due rather to the prophetic *character of the subject-matter*. One illustration of this, which strikes me while I write, I may here give. The unusual word ῥομφαία for a "sword" is used in the Apocalypse sixteen times, but nowhere else in the New Testament, save once, and that in a *prophetic utterance* (Luke ii. 35).

As specimens of the prophetic style in the Old Testament prophets, let any one compare Isaiah xiii., xiv., with the *prose* of the same Isaiah; or the δεινότης of Ezekiel xxvii., xxviii., with the *prose* of the same Ezekiel in such places as xxiii. 21, 22; or our Lord's own style in His terrific denunciations of the "scribes and Pharisees, hypocrites," in Matthew xxiii. 13 to end, and the style of His prophecy of Jerusalem in Matthew xxiv. or Luke xxi., when compared with the inimitable prose of His parting address to the Eleven at the supper table, and the high-priestly prayer with which it closes—

> "Oh, it came o'er my ear like the sweet south
> That breathes upon a bank of violets,
> Stealing, and giving odour,"—

and he will not doubt, I think, that it was the *subject-matter* that gave birth to the *style* of the Apocalypse. In fact, every good writer's style varies with what he writes about. And if proof were wanting that the apocalyptic seer was no stranger to this ability, we need only refer to the pure *prose* of the Epistles of the seven Churches (chap. ii., iii.), in the very midst of which two of the solecisms occur.

3. If the following features of the Apocalypse have any truth in them, the advocates of the *late* date of that book entirely misunderstand it. And as the quarter from which the statement of them comes is entitled to great deference, I must examine it in detail. It is thus expressed by Dr. Westcott, in the Introduction to his great work on the *Gospel of St. John*:—

"The Apocalypse is doctrinally the uniting link between the Synoptists and the Fourth Gospel. It offers the characteristic thoughts of the Fourth Gospel in that form of development which belongs to the earliest apostolic age. It belongs to different historical circumstances, to a different phase of intellectual progress, to a different theological stage, from that of St. John's

Gospel; and yet it is not only harmonious with it in teaching, but in the order of thought it is the necessary germ out of which the Gospel proceeded by a process of life." *

With submission, I venture to say, that there is no such relation between the Apocalypse and the Fourth Gospel as is here described. Harmonious indeed they are in their teaching, but the one is in no sense the germ of the other. The truths common to both are presented historically in the one book, in the other scenically. In the Gospel they appear in their abstract, settled, enduring form, unaffected alike by time and by circumstances; in the Apocalypse they appear in the concrete form, taking their shape from definite circumstances and specific occasions. In the one case, the interest they possess lies wholly in what is of eternal moment; in the other case, it lies in the changing forms which the great struggle between the organised kingdoms of light and of darkness assume in successive ages. With what propriety, then, can it be said that the one is the necessary germ out of which the other proceeds? that the one represents an earlier stage in the development of the same characteristics as the other? In this respect I venture to think that they admit of no comparison.

There is, however, a true, a most important sense which the truths common to both books appear in a less developed form in the one book than in the other. But it is not in the Apocalypse, but in the Fourth Gospel, that that less developed form appears. In that latest Gospel the developed results of God's redeeming love and of Christ's finished work could not possibly appear. But in the Apocalypse they stand out in a form so naked, so rich, so thrilling, as to endear that book to thousands who never

---

* *Gospel of St. John,* "Introd.," p. lxxxiv. (Murray, 1882.)

attempt to sound the depths of its prophetic mysteries. "*I have many things to say unto you,*" were among the last words which the Master addressed to the Eleven before He suffered, "*but ye cannot bear them now. Howbeit when He, the Spirit of truth, is come, He shall guide you into all the truth:* for He shall not speak from Himself; but what things soever He shall hear, these shall He speak: *and He shall declare unto you the things that are to come.* He shall glorify Me: for He shall receive of Mine, and shall declare it unto you. All things whatsoever the Father hath are Mine: therefore said I, that He taketh of Mine, and shall declare it unto you." The best commentary on these words is, first of all, the Acts of the Apostles, from beginning to end; and yet, even there, we find ourselves only in the vestibule of the temple of "the Spirit of truth." Only after the disciples had been formed into churches, needing further instruction by the precious Epistles written to them, do we see how the Spirit had "guided them into all the truth," making them "able to comprehend what is the breadth, and length, and depth, and height, and to know the love of Christ, which passeth knowledge, and be filled with all the fulness of God." So long as the Master was with them, the very language in which such things are expressed in the Epistles would have been unintelligible. But once ascended on high, and the Holy Ghost resting on the Church, the apostles could say to the churches they had gathered, and the churches could understand them when they said, "In whom we have redemption through His blood, the forgiveness of sins, according to the riches of His grace"; and so, in every varied form, in all the Epistles. But the Apocalypse lifts us to even a higher region, giving forth the same truths in strains so exalted as almost to dim the brightness of them everywhere else. There the veil seems to be lifted, and we are ushered into the midst of things invisible and inaudible, with eyes to see and ears to hear. What is elsewhere simply *announced* is here *enacted*; what elsewhere is *said* is here *sung*, sweeping upon the ear in strains celestial. "We love Him (says the beloved disciple), because He first loved us,"—words which will never die upon the lips of any that has ever felt it. But here he rises even above himself, bursting out into song :—" Unto Him that loved us, and washed us from our sins in His own blood, and hath made us kings and priests unto God and His Father, to Him be glory and dominion for ever and ever. Amen." In the Epistle to the Hebrews we read, "To them that look for Him shall He appear the second time without sin unto salvation." Delightful prose, indeed ; but, as if that were too tame, here we seem to see Himself darting through these heavens, to the view of every eye : "Behold, He cometh with clouds ; and every eye shall see Him, and they also which pierced Him ; and all kindreds of the earth shall wail because of Him."

But do I see evidence in this of a later *date* for the Apocalypse ? So far from that, I believe that his Gospel, his Epistles, and the Apocalypse were all written by the last of the apostles in his old age. But, instead of its being " doctrinally the uniting link between the Synoptists and the Fourth Gospel," " the necessary germ out of which by a process of life the Fourth Gospel proceeded " (a view of the subject very wide of the mark, as I humbly think), it is in my view simply the same truths, which in their ripest stage and fullest development appear in the Epistles, lifted up (in the entrancing effect which they produce upon the heart) to the third heaven.

How true this is, grows upon one the farther he advances in the study of the book. When we come to the strictly prophetic part of it, we have at the outset a grand introductory vision in two parts: in the first part (chap. iv.) of God as *Creator*; in the second (chap. v.) of Christ as *Redeemer*; in both cases, however, it is in a language of its own, the significance of which is such that it seems, by its symbols and scenic actions, to compress within a nutshell all that is grandest and richest in every other part of Scripture. In the first part we have "Him that sitteth upon the throne," in whose ears day and night is heard the cry, from one class, of "Holy, holy, holy, Lord God Almighty, which was, and which is, and which is to come," and from another, "Worthy art Thou, our Lord and our God, to receive the glory and the honour and the power: for Thou didst create all things, and because of Thy good will they were, and were created." This is the hymn of *creation*. Now for *redemption*. In the right hand of Him that sat on the throne is seen a book (the book of the Church's fortunes). A challenge is addressed in a loud voice to all creation, for one worthy to open it and reveal its contents, if such could be found. But none answering, the seer weeps much, as if the case were desperate. But he is soon relieved with an assurance which can only be expressed in the angelic language, "Behold, the Lion of the tribe of Judah, the Root of David, hath conquered" (ἐνίκησεν) the right "to open and loose" the seals of this mysterious book. Whereupon, "in the midst of the throne, and in the midst of the living creatures, and in the midst of the elders, I saw a LAMB standing, *as though it had been slain* (in the eternal freshness, the all-atoning virtue, of His precious blood), having seven horns and seven eyes, which are the seven Spirits of God, sent forth into all the earth" (the omnipotence and omniscience of the Spirit in the hands of the enthroned Lamb over the whole earth, to conquer for Himself His inheritance, the uttermost parts of the earth for His possession). This done, the whole ransomed Church (in its twofold character of priests and kings, "the living creatures and the elders"), with their harps and the sweet incense of their deepest emotions, sang that "new song" which will never grow old, "'Thou art worthy to take the book and loose its seals: for Thou wast slain, and didst purchase us with Thy blood out of every tribe, and tongue, and people, and nation; and madest them to be unto our God kings and priests: and they do reign (or shall reign) upon the earth,"—that earth which the fall sold into the hands of the usurper, now cast out. The angels then join in the chorus, but one note is now left out. "Worthy is the Lamb that hath been slain" (not now "for us"); and this is at length taken up by the whole creation, in a fourfold ascription of "blessing, and honour, and glory, and power to Him that sitteth upon the throne, and to the Lamb, for ever and ever"—thus clasping both parts of this incomparable vision in one, while the Church says the "Amen" to "Him that liveth for ever and ever."

But what bearing, it may be asked, has this upon the question of *date*? To me it suggests this question: Is it natural to suppose that a book presenting the most exalted conceptions of the glory and majesty of the Eternal, with the ripest and richest expressions of the Person and work of Christ, and both these breaking upon our ear in strains of celestial music, was written so much earlier than the Fourth Gospel that "it belongs to the earliest apostolic age"? For myself, I cannot believe it.

But this becomes more difficult to believe as we advance in the visions of the book. We have seen how the two central objects, "GOD and THE LAMB," stand out together in the great introductory vision. In chap. vii. the seer beholds "standing before the throne, and before the Lamb, a great multitude which no man could number, out of every nation, and of all tribes, and peoples, and tongues, arrayed in white robes, and palms in their hands, and crying with a loud voice, Salvation unto our God, which sitteth on the throne (the *Source* of it), and unto the Lamb (the mediatorial *Channel* of it)"; while round about them stood all the angels, who fall on their faces, and worship Him that liveth for ever and ever. But what is the secret of those "white robes" and their right to "stand" before the throne, since from the very face of Him that sitteth upon it "the earth and the heaven fled away" at the last judgment, "and there was found no place for them"? They had come out of the great tribulation, and had "*washed their robes, and made them white, in the blood of the Lamb.* Therefore are they before the throne of God," etc. Does such language read as if it "belonged to the earliest apostolic age"? That is not my reading of the New Testament.

A word on the surpassing strains of the two last chapters. When one reads in the *Pilgrim's Progress*, how, when winding up his immortal allegory, the author's language sounds like the music of heaven, he is ready to say, Was ever such a finish given to any story? But whence did he draw his inspiration? From this book; but for whose closing chapters—fit close to the inspired volume itself— we are safe to say such language could never have been penned. Yet this, we are to believe, "belongs to the earliest apostolic age"!

These, however, are but great generalities. To me there are certain specific characteristics of this book which speak for anything but an *early* date.

1. The Church of God under the old dispensation was one undivided whole, existing only in " the Lord's land," and its central seat was on Mount Zion, in the Tabernacle and Temple. Accordingly the golden candlestick, or lampstand, was *one*, and the seven branches of it, when all lighted up, gave light to the whole interior. But in the Apocalypse, at the very outset, the seer beheld, not one, but "*seven* golden candlesticks" or lamp-stands; and as these represented the seven distinct churches, there is here announced a complete ecclesiastical revolution—the Church of God, in its external framework, no longer one, but broken up into sections corresponding with the geographical divisions of its members. Our Lord gives a very distinct intimation that such a division of "His sheep" was at hand. Speaking of His true disciples, who at that time were all Jews, He says : " I lay down My life for the sheep. And *other sheep I have, which are not of this fold: them also I must bring*, and they shall hear My voice : *and they shall become one flock and one Shepherd*" (John x. 16). The A.V., by translating "one *fold*," expresses the precise theory of the Church of Rome, that the whole Church of Christ should be within one pale. But since the seven churches of Asia were in all outward respects as distinct from one another in their corporate existence and internal condition as were the localities in which they were placed, so we must hold the teaching of the Apocalypse to be, that the Church of Christ is intended to consist of as many distinct and independent branches as the different localities (or perhaps impossible combinations) in which they find themselves.

That such a conception could not have found a place in a book written "in the earliest apostolic age," and in so distracted a time as on the eve of the destruction of Jerusalem, I do not say. But in my judgment it clearly belongs more naturally to a later stage and a more settled state of things in the development of the Church of Christ. But this brings me to the Epistles to the seven churches.

2. I am not disposed to make too much of the degenerate character of some of these seven churches, especially of the last one, Laodicæa, as an evidence of the *late* date at which they must have been written. But taken along with other arguments in the same direction, this degeneracy is certainly noteworthy. Take the first of these Epistles—to *Ephesus*. Three years or more after this church sprang up, in a city steeped in a gorgeous and witching idolatry, its spiritual father addressed them through its assembled elders at Miletus (Acts xx.); but though he had to warn them against an influx of self-seeking teachers, and false brethren among themselves, the Epistle which he wrote to them about four years after that, so far from showing that they had sensibly declined, teems with evidence implying rather a steady condition. But when the Master addresses this church in the Apocalypse—within four or five years only after that, if the Neronic date is adopted—it had so sunk that, should it not repent, the removal of its candlestick, or its extinction as a church, would follow. The church of *Sardis* "had a name to live"—a reputation among the churches for being full of spiritual life—"but was dead": the life they had having died down, and they were living upon their reputation. The white raiment given them at their conversion (compare Zech. iii. 4) had been so ill kept that but "a few names" could be found who "had not defiled" them. As for the *Laodicæans'* condition, it was so loathsome in the pure eye of its exalted Lord, that He likens it to food which one is fain

to vomit up. Does this look like churches only a few years in existence?

3. THE LAMB, as a proper name, is never applied to Christ by any New Testament writer save John, and even by him nowhere in his Gospel nor in his Epistles. For though it occurs twice in his Gospel, he is there reporting an exclamation of the Baptist (John i. 29, 36). But when we come to the Apocalypse, we find it no fewer than *twenty-eight* times. In fact, the constant recurrence of this remarkable epithet is a special characteristic of the book; and if the reader will attentively examine this phrase as it occurs in chap. v. he will come, I think, to this conclusion: that while it presents the great central truth of the atonement in no more fully developed form than in the Pauline and Petrine Epistles, it is a form which, if it had struck upon the ear of the Church in the time of these apostles, would have become the *current coin* of its phraseology—one of those household words which could not fail to crop out here and there, if only for variety, in their writings. But since we find it nowhere but in this book, it is to me no slight evidence that it was not in existence in their day.

4. "*The books of life*" is a phrase used only once elsewhere in the New Testament (Phil. iv.), as a record of *names*, the names of the righteous, which God is supposed to keep; we find it as early as the days of Moses (Exod. xxxii. 32). The psalmist catches it up (Ps. lxix. 28; cxxxix. 16). In Malachi (iii. 16) it is said, of a time of deep religious declension, that "a book of remembrance was written before Him for them that feared the Lord, and that thought upon His name." But in Daniel, *the apocalyptic book of the Old Testament*, besides occurring in the definite form, "the book" (xii. 1), we have the phrase in a quite

distinct form—that of "*books*" in the plural number. The scene in which it occurs is a scene of "the last judgment"; but it is not of individual *men*, but of *nations* in their corporate capacity, and therefore *here on earth*. It is the judgment of the four kingdoms previously specified, the oppressors of the Church, together with one terrible form of the fourth one. "Thrones" of judgment being "placed," One that was the "Ancient of days did sit" for judgment in terrible majesty, surrounded by myriads of angels; the judgment was set, "and THE BOOKS were opened." This was no record of *names*, but of the *deeds* of those kingdoms for which they were to be condemned, and their kingdoms were to give place to that of the "Son of man," whose kingdom is an everlasting kingdom. Now, observe how this double conception of "the *book*" (of names) and "the *books*" (of deeds) is taken up in our New Testament Apocalypse. Four times it comes before us; but I begin with the place where they both appear in a very definite and most wary form (chap. xx. 11, 12, 15). "And I saw a great white throne, and Him that sat upon it, from whose face the earth and heaven fled away, and there was found no place for them. And I saw the dead, the great and the small, standing before the throne; and *books* were opened, and *another book* was opened, which is *the book of life*: and the dead were judged out of those things which were written in *the books*, according to their works. . . . And if any one was not written in the book, he was cast into the lake of fire." The sublime idea conveyed by this artistic distinction between "the book" and "the books" is that the judgment proceeded exclusively upon "their works," as recorded in "the books"; but that, this done, "that other book was opened," from which it appeared that this decision, both upon the righteous and the wicked, had been recorded in that book "from the foundation of the world" —the names of those adjudged by "the books" to "eternal life" being exclusively found there, while the absence of the names of all others ("if any one was not found written" there) expressed negatively what would be found to be their due. Thus two characteristics of this book come out: its being the book of those "ordained to eternal life" (Acts xiii. 48), and its having been written "before the foundation of the world." Another characteristic of vital moment is, that it is "the book of life of the Lamb that was slain"—that is, specifically in His sacrificial character; teaching this great truth, that the names found in this book were written there solely in virtue of their connexion with His atoning death, eventually to take place. And so it is called "the Lamb's book of life." *

What conclusion now, as to the *date* of the Apocalypse, do I draw from these facts? Decidedly this: that the whole conception of "the book of life," and "the books" out of which the dead will be judged, has advanced progressively in the outcome of Divine revelation, and that, appearing only in its fullest, most artistic, and most speaking form, it proclaims its place in the order of time to

---

* It is a thousand pities, I think, that both the Authorised and Revised Versions punctuate chap. xiii. 8 thus: "The book of life of the Lamb slain from the foundation of the world." No such idea as this, that Christ was crucified before the foundation of the world, is anywhere else to be found in the New Testament; and if any one will compare the same idea of chap. xiii. 8 as it is repeated in xvii. 8, where only the writing of their names from the foundation of the world (not His being slain from that time) is mentioned, he will see, I think, that the following is the proper punctuation of the verse: "All whose names are not written from the foundation of the world in the book of life of the Lamb that hath been slain" (as in the margin of the R.V.).

be in this book, as the fitting close of all revealed truth. But—

5. What shall I say of the almost countless number of phrases peculiar to this book, but full of pregnancy? Take the seven epistles—"the first love" of one church "left," and "the last works" of another church "more than the first"; the burning eyes of their exalted Head having "a few things" against two of the churches otherwise praised; one church commended for having "a little strength," and not "denying His name," "the second death," and so on. Then, in the prophetic part, the central position given to the symbols of the living creatures and the elders (because representing the redeemed, as is evident from chap. v. 9), while *outside* of them and surrounding them are the angels, who also ascribe worthiness to the Lamb that was slain, but do not say "for us," as in verse 9). And above all, the constant conjunction of "Him that sitteth upon the throne and the Lamb," the one as the fountain whence flows all salvation, the other as the channel through which it all flows to men. In chap. v. the relative position of each respectively, and their absolute oneness in the work of redemption (as in John xiv. 7, 9—11, 23; xvi. 15; xvii. 21), stand boldly out; but in chap. vi. 16 we have them awfully associated in "the wrath of Him that sitteth upon the throne, and *the wrath of the Lamb*." In a word, that peculiar name given to the enemy of souls, suggested by his occupation, "the accuser of the brethren," who "accuseth them before our God day and night." In fact, the whole book teems with unique epithets and phrases, and symbolic arrangements, suggestive of the "unsearchable riches" of that scheme of salvation which, while expressed it is true in fully developed forms in the apostolic Epistles, appears in this book as if the seer had been

instructed to take us, not into the sanctuary only, but into the holy of holies.

In view of all this, can it be said that this book reads like "the connecting link between the Synoptists and the Fourth Gospel," and "that form of development which belongs to the earliest apostolic age"? Let the reader judge.

III.

DESIGN OF THE APOCALYPSE.*

THERE are but two possible theories of what the Apocalypse was written for. It is either essentially *predictive* or purely *descriptive*. Its proper subject-matter is either *events* or *ideas*. In the one case, its purpose is to foreshadow the future fortunes of the Church, at successive epochs of its history; in the other case, to set forth, in symbolic scenes and dramatic movements, the great *principles* that have been struggling for the mastery in all ages and in different forms—light and darkness, good and evil, the so-called *World-Power*, whether Egypt or Babylon, Pagan or Papal, in hostility to the kingdom of God.

What I propose in this paper is, to examine the claims of the non-predictive, or purely descriptive theory. And I will let its advocates themselves explain it. For this purpose I select the two most recent English expositors of this book. In the *Speaker's Commentary* the late Archdeacon Lee thus writes:

"The book of Revelation (says Ebrard) does not contain passages of contingent events, but certain warnings and consolatory prophecies concerning the great leading *forces*† which make their appearance in the conflict between Christ and the enemy. So full are its contents, that every one may learn more

---
* From the *Expositor* (1889), vol. x. (3rd series), pp. 444-56.
† The italics are mine.

against what disguises of the serpent one has to guard himself, and how the afflicted Church at all times receives its measure of comfort and consolation. The imagery of the book (continues Dr. Lee) naturally describes, in accordance with the whole spirit of prophecy, the various conditions of the kingdom of God on earth during its consecutive struggles against the prince of this world. . . . The spiritual application is never exhausted, but merely receives additional illustration as time runs on" (*Introd.*).*

Hear now Professor Milligan :

" It is a book which deals with *principles* † rather than particular *events*. The same remark, indeed, is applicable to all the prophetic books of Scripture ; for these are *for the most part* occupied with principles that are generally, even universally, fulfilling themselves in human life. . . . They are proclamations of eternal truths—of the sovereignty of God, of His superintendence of the world, of His approbation of good, of His hatred of evil, of the fact that, notwithstanding all the apparent anomalies around us, He is conducting to final triumph His own plan for the establishment of His righteous and perfect kingdom. It is well therefore that prophecy should be uttered *to a large extent* in general language. The men of one age see it fulfilled in what passes around them; the men of another age do the same. The struggle between the principles of good and evil marks all time. It returns in every age, and God is always the same God of judgment." ‡

To do justice to this theory is far from easy, from the vague way in which its advocates express themselves. But one or two things seem obvious.

---

* Dr. Lee calls this the *spiritual* view of the book; but what his own principle of interpretation is it is difficult to discover, for his exposition consists of little else than a *catena* of interpretations which he himself does not accept.

† The italics in this extract are mine.

‡ *Popular Commentary on the New Testament* (Dr. Schaff's) : vol. iv., " Revelation."

1. Was this book written for no other purpose than to proclaim the sovereignty of God, His superintendence of the world, His approbation of good and hatred of evil, and how, in spite of anomalies, He is conducting to final triumph His own plan for establishing a righteous kingdom? Were these first principles, these elementary truths, of all revealed religion so obscurely expressed and so insufficiently enforced in other parts of Scripture, that it needed a book of such complicated structure and such extreme difficulty of interpretation, to make them clearer and more impressive? Why, they are themselves infinitely plainer than the book which we are told was written to enforce them. Whatever may be thought of other theories, this at least will never do.

2. It is scarcely self-consistent. Its advocates seem to oscillate between the predictive and non-predictive view of its contents. At one time we are told not to look for actual history in it; but anon they say it "deals *rather* with principles than particular events. The same remark," adds Dr. Milligan, "applies to all the prophetic books of Scripture, which *for the most part* are occupied with principles. It is well therefore that prophecy should be uttered *to a large extent* in general language." Now, what is the use of this constant guarding against looking for "historical events" in prophecy? The question is, Are there *any* such? That there are, your own language admits; for you say it is only "for the most part" and "to a large extent" that it deals in "general principles," and that it deals "rather" in these—of course implying that it *does* deal, to some extent, in "historical events." And yet we are warned not to look in prophecy for such events. The one question clearly should be, What *is* and what is *not* predictive? That is a purely exegetical question; and, tried

by this test, it is hard to see how any other than a predictive design this book can possibly have. The very first words of the book speak for themselves: "The book of the Revelation of Jesus Christ, which God gave unto Him, to shew unto His servants the things which must shortly come to pass"; and a very unusual blessing is pronounced, and in the next words, upon "him that readeth, and them that hear the words of this prophecy, and keep those things which are written therein: *for the time is at hand.*" If this does not mean that definite historical *events* were about to happen, for which the churches were warned to look, what can we make of such language? But is not our Lord's prophecy of the destruction of Jerusalem full of concrete historical predictions? And the apostle's prophecy of "the man of sin"—whatever it may mean—does that not bristle with concrete historical predictions? To what purpose, then, is it to say that prophecy deals "for the most part" with general principles? If the Apocalypse is *not* such a book, it is entirely beside the mark.

3. This theory, in its systematic form, is, so far as I know, entirely novel. I am not aware of one commentary on the Apocalypse constructed on this principle until towards the close of the last and early in the present century, when a tide of anti-supernaturalism set in upon the Church, especially in Germany, begetting a rationalistic criticism that explained away both miracles and prophecy. But if it be asked how to explain the rise of this novel theory among believing expositors, I ascribe it to despair of finding in history any events to correspond with the predictions, suggesting at length the question, What if it was never meant to predict historical events at all? May not its sole design be to hold forth in bold relief, and under the guise of old historic foes of the kingdom of God—Egypt, Babylon,

Jerusalem—the ever-recurring assaults upon the kingdom of darkness?

The ablest and most ingenious exposition of this scheme of interpretation is that of the late Dr. Arnold, in his two sermons on the interpretation of prophecy.* Since his time the anti-predictive theory of apocalyptic interpretation seems to have taken hold of a class of English interpreters of both Old and New Testament prophetic Scripture. To bring this theory to the test I know not any better way than to try it on the commentaries already referred to. To Dr. Lee I need not refer, because, as already said, his exposition of the prophetic part of the book gives no clear indication of how his theory comes out at all. But my esteemed friend Dr. Milligan is a pleasant contrast to this, his exposition being rigidly exegetical from first to last—the text and the symbols being explained with elaborate minuteness, and adhering with admirable fidelity to what he takes to be the one object of the book, to explain and illustrate great "general principles"—not to predict at all.

Thus far I had written two years ago, when, on receiving Dr. Dods' *Introduction to the New Testament*,† I found Dr. Milligan's theory rejected in terms even more sweeping:

"A still more effectual evasion ‡ of the difficulties attaching to any historical interpretation, whether Præterist, Futurist, or continuously Historical, is suggested by Dr. Milligan, who proposes that we should read the book as a representation of ideas rather than events. It embraces, he thinks, the whole period of the

---

* *Sermons on the Interpretation of Scripture*, 3rd ed., 1878, pp. 333-94.

† "Theological Educator" series, edited by Rev. W. R. Nicoll. (Hodder & Stoughton, 1888.)

‡ Not of course intentional, Dr. Dods would admit.

Christian dispensation; but within this period it sets before the reader the action of great principles, and not special incidents. It is meant to impress the reader with the idea that many years of judgment, of trial, of victory must pass over the Church before the end comes. The end, indeed, is spoken of as near; but this results from the impression which could not but be received by the early Church, that now that Christ had actually come the end was virtually present. 'The book thus becomes to us, not a history of either early or mediæval or last events, written of before they happened, but a spring of elevated encouragement and holy joy to Christians in every age.' It exhibits the Church of Christ in its conflict, preservation, and victory; and it sees these through the forms and in the colours presented to the writer's imagination by what he himself had seen and experienced, and by his knowledge of the Old Testament and of our Lord's discourses. It is not a political pamphlet disguised, but a vision of the Church's necessary fortunes as the body of her Lord, and His representative on earth. Babylon therefore is not pagan Rome, but the apostate Church of all ages, described in a highly elaborated picture, of which the outlines had already been drawn by the prophets. This system of interpretation has its attractions, but is certainly (1) out of keeping with the general purpose of apocalyptic literature, and (2) fails to present a sufficient motive for its composition, and (3) a sufficiently definite guide through its intricacies" (pp. 243, 244).

Of the three objections to which I have attached figures, I have dealt pretty fully with the second and third. But while it is true (according to the first) that it is out of keeping—indeed, glaringly so—with the general purpose of apocalyptic literature, I must guard against the abuse to which that phrase is liable.

Of the prophetical books of Scripture, those of Daniel in the Old Testament and Revelation in the New differ widely from all the rest. In both books the subject treated of is the kingdom of God oppressed by hostile worldly powers; in both books successive periods in the history of this struggle are definitely though symbolically predicted; in both the protracted character of the struggle, as well as the final overthrow of these hostile powers and the triumphant establishment of the kingdom of God, are set forth to cheer the hearts of the faithful; while in the latter book the chronology of the conflict in its successive stages is specified with a marvellous minuteness of detail, perhaps befitting the last word of Divine revelation. There is nothing in the least like this in the other prophetical books; and this characteristic is adequately expressed by the word "apocalyptic."[1]

But such hold did this feature of the book of Daniel take upon the Jewish mind after the captivity, groaning under successive oppressions, that it gave birth to productions of the same character, holding forth the expected redemption according unto the conception of their several writers; and so fascinating was this kind of literature that, even after the New Testament "Revelation" appeared, similar writings—or mixtures, rather, of it and Jewish works of this kind—were sent forth. The consequence of this has been, that modern critics have come to mass up all such writings, from Daniel to Revelation and onwards, under the common name of "apocalyptic literature." I cannot assent to this. Any one who compares the Book of Daniel of the Old Testament and the Apocalypse of the New must see at a glance that they stand or fall together; that the New Testament Apocalypse is expressly intended as a sequel to and completion of the disclosures in Daniel about the four empires: so that if the Book of Daniel is not a genuine and authentic work, neither is the New Testament Apocalypse; whereas if this last book of the New Testament be indeed "the Revelation of Jesus Christ which God gave unto Him," to forewarn the Church of coming events, so also is its *prodromus*, the Book of Daniel. In fact, nothing could express the connexion between the two books more neatly than the phrase of Mede, that Daniel is *Apocalypsis contracta*, while the Apocalypse is *Daniel protracta*. To mass up these two books therefore with that heap of writings in imitation of them called "apocalyptic literature," ranging from the merest rubbish up to those of more or less pretensions to respectability, is not to be endured.

(The best known of these are the books called "Second Esdras" in our English Apocrypha and the "Book of Enoch." A pretty full account of both will be found in the *Encyclopædia Britannica*, 9th ed., art. "Apocalyptic Literature," especially of the Book of Enoch. For the English reader the most serviceable version of it is one made by Professor Schodde of Ohio.—Andover, 1882.)

But what is to be said to the critics of the modern school, who freely admit that historical events, and not mere ideas, are the proper subject of this book, and insist therefore that "all interpretation not strictly historical must be excluded"?* But so far from being predictive in any legitimate sense of the word, they find them all living in the near distance to that of the writer, and some of them in the course of actual occurrence in his time, requiring therefore no higher inspiration than keen insight into the signs of the times. So confident are such critics that they have at length got the true "key" to the Apocalypse in their hands, that they are bold enough to affirm that "the matter of the book is neither obscure nor mysterious," and "without being paradoxical, we may affirm that the Apocalypse is the most intelligible book of the New Testament!"† With

---

* *Encyclopædia Britannica* (9th edition), art. "Revelation," by Professor Harnack.

† *Ibid.*

these critics, everything exegetical in the interpretation of this book is "settled" and "beyond dispute." This is not the stage of our subject at which we can examine their interpretations in detail, but when we come to "The Structure of the Apocalypse," it will soon be seen that their "key," at least, will not do much to help us.

INTRODUCTION.

ADDENDUM.

SIR WILLIAM HAMILTON'S ATTACK ON THE APOCALYPSE.

SIR WILLIAM HAMILTON'S ATTACK ON THE APOCALYPSE.

[Sir William Hamilton, Professor of Logic and Metaphysics in the University of Edinburgh, was an accomplished classical scholar, a profound metaphysician, and an omnivorous reader—equally at home in German, French, and Italian. In fact, what was said of Professor Whewell was equally applicable to him,—that knowledge was his forte and omniscience his foible. His literary reputation was made by his elaborate articles in the *Edinburgh Review*, when it was in its palmiest days. They were full of literary interest, and some of them were treatises, and the most valuable of them were re-issued by himself in a massive 8vo volume of nearly 900 closely printed pages, entitled, "Discussions in Philosophy and Religion," etc. In temper he was too much of an Ishmaelite: he could ill brook contradiction; and his pen, in replying to an opponent, was too often dipped in gall. The attack, to which the following was a reply, occurs incidentally in an article on "The Right of Dissenters to Admission to the English Universities," and my only reason for reprinting it here is that it gives a number of curious and interesting biographical facts which it took me a good deal of time to hunt out, and which should not go quite out of sight.]

"How could Mr. Pearson make any opinion touching the Apocalypse matter of crimination against Semler and Eichhorn? Is he unaware that the most learned and intelligent of Protestant [of Calvinist]* divines have

---

* The bracketed words were added by Sir William in a subsequent edition.

almost all doubted or denied the canonicity of the Revelation? The following rise first to our recollection. Erasmus, who may in part be claimed by the Reformation, doubted its authenticity. Calvin and Beza denounced the book as unintelligible, and prohibited the pastors of Geneva from all attempt at interpretation; for which they were applauded by Joseph Scaliger, Isaac Casaubon, [and our countryman Morus, to say nothing of Bodinus, etc.] Joseph Scaliger [of the learned the most learned], rejecting also the Epistle of James, did not believe the Apocalypse to be the writing of St. John, and allowed only two chapters to be comprehensible; while Dr. South [a great Anglican authority] scrupled not to pronounce it a book (we quote from memory) that either found a man mad or left him so."—(*Discussions in Philosophy, etc.*, p. 506.)*

This assault upon the Apocalypse was altogether gratuitous. Sir William had a sufficiently good case against his opponent—"the Rev. George Pearson, Christian Advocate in the University of Cambridge"—who, to show "the danger of abrogating the religious tests and subscriptions which are at present required from persons proceeding to degrees in the universities," had launched out in his pamphlet into a crude enough statement of the errors which have been broached by German professors regarding the books of the Bible; the whole being wound up with the following sentence on the Apocalypse:—"Eichhorn

---

* The words enclosed in brackets have been added (in 1852) to the original statement as it appeared in the *Edinburgh Review*; showing that, though the paragraph has been carefully *revised*, it has in no respect been *corrected*, and that Sir William Hamilton, after the lapse of nearly twenty years, is in theological matters as careless of his reputation for accuracy as ever he was.

pronounces the Revelations to be a drama, representing the fall of Judaism and Paganism; while Semler condemned it entirely as the work of a fanatic." If this statement deserved any notice at all, as bearing on the test question, which we hardly think it did, it might have been enough to say, that the man who thus expresses what Semler has said on the Apocalypse must have taken his information at second hand; that to mix up Semler, as adverse to the canonical authority of the Apocalypse, with Eichhorn, who both held its canonical authority and published an exposition of it, was more like a special pleader than an impartial writer; and that, even admitting the case to be as stated, it was not sufficient to support the conclusions founded upon it. This would have sufficed for our author's argument, and thus far we could have gone heartily along with him. But so favourable an opportunity of showing his Apocalyptic lore, of having a thrust at the book itself, and relieving himself of his superfluous Ishmaelism, was not to be lost. Hence this strange paragraph on the Apocalypse, which we are now to take up in detail.

1. "How," asks Sir William, "could Mr. Pearson make *any* opinion touching the Apocalypse matter of crimination against Semler and Eichhorn? Is he unaware that *the most learned and intelligent of Protestant, of Calvinist divines, have almost all doubted or denied the canonicity of the Revelation?*"

On this assertion we make the following remarks:— *First*, It is not true. We challenge Sir William Hamilton or any man to prove it. *Second*, Though all the authorities adduced had questioned or denied "the canonicity of the Revelation," they are no proper representatives of "almost all the most learned and intelligent Protestant and Calvinist divines." Sir William's list of authorities is an absurd one for his point; for, with the exception of Calvin and Beza (of Erasmus we shall speak presently), not one of those named would naturally be thought of as an authority in a question of this nature, a question which lay out of the region of their special studies. Calvin and Beza, if correctly reported, are entitled to great weight; but even they do not quite stand for all the "learning and intelligence of Protestant and Calvinist" theology on this question. But, *Third*, Even Sir William's witnesses must, on his own showing, be put out of court, with the single exception of Erasmus. For, whereas the thing "doubted or denied by almost all the most learned and intelligent of Protestant, of Calvinist divines," is, according to our author, the *canonicity* of the Revelation, he brings in his witnesses to speak to quite another point—namely, whether people are able to *understand* this confessedly mysterious book. On that point, some strong, and, as we think, rash things have been said by many who yet never questioned the canonicity of the book. When Sir William, therefore, introduces them to us, to inform us that they could make nothing of the Apocalypse, we just walk them out again, as useless for his purpose. One, indeed—Joseph Scaliger—is made to speak to the *authorship* of the book, which comes nearer to Sir William's point. But even this does not settle the question of its canonicity; for though, if the beloved disciple was the writer, its canonicity is of course established, every one knows that some eminent critics have ascribed this book to another John, who nevertheless maintain its canonicity. Thus Sir William's own witnesses are made by himself to disappear from the stage, with the single exception of Erasmus, of whom he can only say, that he "may in part be claimed by the Reformation," having been neither a "Calvinist nor a Protestant divine."

2. What Sir William says of *Erasmus* is correct enough, that he "doubted the authenticity" of the Apocalypse. But the value of this doubt remains to be investigated. Here we shall simply translate from Beza's *Prolegomena* to the Apocalypse:—" As some have long since doubted the authority of this book, I shall first briefly demolish the arguments usually employed on that side, and state my own views. I will give the arguments as they have been studiously and industriously collected by Erasmus, *whose own judgment, however, on this as on many other points, seems to me so wavering, that one cannot discover what he really thought,* except that he seemed inclined at length to believe that *some kind of authority belonged to this book*, though not what attaches to the books which have been received without controversy." With this we leave Sir William's reference to be taken for what it is worth. On a question of mere criticism, the opinion of Erasmus is entitled to the greatest weight. But those who read his arguments, as stated by himself, and as reported by Beza, will see at once that other considerations, quite as much or rather more than critical, influenced Erasmus in his doubts about this book; and we do him no wrong when we say, that on these *other* considerations Erasmus is entitled to no more weight than any other student of the New Testament.

3. "Calvin and Beza denounced the book as unintelligible, and prohibited the pastors of Geneva from all attempt at interpretation." This we have no hesitation in pronouncing a scandalous statement. It is notorious that both Calvin and Beza held the Apocalypse to be a canonical book. If Sir William did not know that, he should have let the subject alone till he was better informed; but if he did, it was most improper and offensive to say, that "Calvin and

Beza *denounced* the Apocalypse." Even though neither of them had thought it could be explained, we may be quite certain, from their known reverence for all that they held to be the word of God, that they would never "denounce" it, in any legitimate sense of that term; and, therefore, we charge Sir William with selecting obnoxious phraseology, on purpose to create a prejudice against the Apocalypse through the aid of two of the greatest names in the Reformed Church. But, further, we challenge Sir William Hamilton to produce, from the writings of either Calvin or Beza, a particle of evidence in support of his assertion. We do not refer here to the absurd statement about the prohibition issued by them to the Genevese pastors. That, we suppose, will be fairly given up as a flourish of trumpets. But we mean their denouncing, or even pronouncing it *unintelligible*. As to Beza, not to speak of his running explanatory notes upon it, which are nearly as many and as long as on any other book of the New Testament, the following words from his *Prolegomena* may suffice to put Sir William Hamilton to shame. After repelling the objections to its canonical authority and apostolic authorship, he says :—
"As to the book itself, although I confess myself one to whom these mysteries are very obscure, yet, when I observe the name of the prophetic Spirit everywhere conspicuous, and perceive not the traces merely, but the sentiments, and even the very words, of the ancient prophets in this book; when I behold throughout clear and most magnificent acknowledgments both of the divinity of Christ and of our redemption; when, in fine, *of the predictions which it contains, some have been manifestly fulfilled, as, for example, those relating to the destruction of the Asiatic churches, and to the kingdom of that harlot that sitteth upon the seven hills, I come to this conclusion, that it was the design of the Holy*

Ghost to collect into THIS MOST PRECIOUS BOOK *such of the predictions of the ancient prophets as remained to be fulfilled after the coming of Christ, and to add to these such others as he deemed to be of importance to us.* Very great obscurity, I acknowledge, there is in them; *but this is nothing new in the writings of the prophets, and especially Ezekiel.* Further, *it is a shame that, engrossed with our own private affairs, we do not study these matters more attentively, and watch those daily evolutions of the providence of God in the administration of His Church.* In a word, the Lord, in His infinite wisdom, has tempered the light of the prophecies to what He foresaw it would be for the good of the Church to know. *It remains, therefore, that men should search these mysteries of a holy God, so far as it is permitted and profitable, with godly fear;* but, at the same time, that all, both those who comprehend, and those who comprehend not, the divine mysteries contained in this Book, should rather *adore them*, than, as some do, either DERIDE THEM on the one hand, or, on the other, pollute them with fanatical comments."

So much for Beza. As to Calvin, any one even moderately acquainted with his "Institutes" can rebut Sir William's statement for himself. But we shall require to return to this ere we have done with Sir William's next assertion, which, as it contains more misstatements than lines, and nearly as many as words, it will be necessary to take up piecemeal.

4. "For which,"—that is, for "denouncing the Apocalypse as unintelligible, and prohibiting the pastors of Geneva from all attempt at interpretation,"—"they" (Calvin and Beza) "were applauded by *Joseph Scaliger.*"

We are sorry to be obliged to say that this is false. As to Beza, we challenge Sir William Hamilton to prove that

Scaliger applauded him for anything whatever touching the Apocalypse. The truth seems to be that Sir William, eager to communicate that important piece of information about *Calvin* and the Apocalypse, which every reader of Bishop Newton's " Dissertations on the Prophecies " knows perfectly well, thought he could not be far wrong in letting *Beza* divide the honour with him. These reformers were such Siamese twins that anything which the one did, or was said to have done, might safely enough be ascribed to the other, which would give him the benefit of two names on his side instead of one. Or, perhaps, our author was so accustomed to associate the one with the other that, having written " Calvin," the words " and Beza " seemed a necessary adjunct, and slipped from his pen quite unconsciously. Be this as it may, until he shall produce his authority for a statement so directly in the teeth of the extract we have given from Beza himself, we must take the liberty of pronouncing it a calumnious fiction.

But, even as respects Calvin, we challenge Sir William Hamilton to prove that Scaliger applauded him for prohibiting the interpretation of the Apocalypse,—applauded him for denouncing the Apocalypse as unintelligible,—or even applauded him for saying that he himself did not understand the Apocalypse. All that Scaliger is ever alleged to have said is, that Calvin showed his wisdom in *not commenting on the Apocalypse.* Now, admitting that Scaliger said this—which we do not, and that the statement is entitled to all the weight which the prodigious learning of Joseph Scaliger might be supposed to lend to it—the reverse of which we shall show to be the case—what does it prove against the Apocalypse ? Why, nothing but that Calvin, conscious that he was not able to throw that light upon the Apocalypse which he had done upon the other books of Scripture on which he commented, was wise enough not to meddle with it. But that Calvin took credit to himself on this score, by no means follows from the mere fact of his letting it alone. He did not comment on Judges, nor on Ruth, nor on Samuel, nor on Kings, nor on Esther, nor on Nehemiah, nor on Ezra, nor on Proverbs, nor on Ecclesiastes, nor on the Song of Solomon. On the other hand, he appeals to the Apocalypse, not merely in illustration, but in proof of important doctrines repeatedly in his "Institutes," just as he does to other books of canonical Scripture. And as Whitby and other commentators have expressly told us that diffidence of their own judgment in the interpretation of prophetic mysteries, and not any wish to disparage that blessed book of Scripture, was the sole reason of their not commenting on the Apocalypse, why should not Calvin, who so often does honour to that book, have the benefit of the same explanation? If so, then Scaliger's statement, supposing he uttered it, and allowing it all the deference which can be claimed for it, makes not one feather's weight against the Apocalypse.

But now, *did* Scaliger say what is ascribed to him about Calvin and the Apocalypse? and, if so, is it worth a moment's attention? It has been so often referred to as a thing undoubted and entitled to some weight—from Bishop Newton down to Dr. Henry of Berlin, in his "Life of Calvin"*—that it may be worth while to reduce it to its proper dimensions.

Be it observed, then, that in Scaliger's own published writings there is not a word on the subject. The statement

---

* *Leben Johann Calvins, von* PAUL HENRY, I. xv. 347, 348. (Hamb. 1835.)

in question occurs in a little book entitled *Scaligeriana*,*
a collection of his casual observations, or table-talk, on
books, authors, and literature in general, taken down, it is
said, from his own lips, and afterwards published. How
far we can rely on these reported observations as having
actually proceeded from Scaliger's lips, or at least in the
very form in which they are there given, we shall leave our
readers to judge from the following account of its origin and
history, which, as that of Hallam is inaccurate,† we shall
take from an old French work, whose circumstantial narrative carries internal evidence of substantial truth :—Two
brothers of the name of De Vassan, French Protestants,
went to Leyden to study, had daily access to Scaliger, and
took down everything curious which fell from his lips. On
returning home, they turned Papists, and made over their
collections to two persons of the name of Du Puy, who
handed them to one Serrau, who left a copy to his son
Isaac, who gave it to M. Daillé—son of the great Daillé
—who, for his own convenience, arranged the articles in
alphabetical order : from him Isaac Vossius got hold of it
when in Paris, and had it published at the Hague, under
the title of "*Scaligeriana.*" ‡ Three years after this there
came out another "*Scaligerana,*" consisting of the collections of a Dr. Vertunien, taken down by him while physician
to two gentlemen with whom Scaliger then lived. These
papers, after the writer's death, had lain in obscurity for
many years, when they were purchased by an advocate of
the name of Sigogne, and published, with a preface and
notes by M. Le Fevre, under the name of " Scaligerana

---

* *Scaligeriana : sive Excerpta ex ore* JOSEPHI SCALIGERI, *per F.F.P.P.*
The copy before us is the second edition, published at the Hague,
12mo, 1668.

† "Lit. of Eur." i. 510.   ‡ See note *.

Prima"—by way of asserting their priority to the other collections which had got the start of it in publication, which other collections it reprinted as an appendix, under the title of "Scaligerana *Secunda,*" as entitled only to the second place, though first issued. This last is a mixture of French and Latin, while the other is all Latin.* Such is the book from which this famous statement is taken.

But our readers may now wish to have the statement itself. It occurs under the head "Calvin," where, after praising his commentary on Daniel and his "Institutes," he is made to say, in French, "*Calvin did very well to write nothing on the Apocalypse.*" After a few additional remarks in the same commendatory strain, the French stops, and the Latin begins, which makes it doubtful whether what follows, though printed in the same paragraph, was not dropped on some other occasion. "The Papists run down Calvin, only because they see his admirable genius and their own inability to come up to him as an interpreter of Scripture. Oh, how well Calvin hits the sense of the prophets! None better. The genius and judgment of Calvin were of the first order. *He was wise not to write on the Apocalypse.*"†

---

* Baillet's "Jugemens des Savans sur les principaux Ouvrages des Auteurs," Monnoyes edition (13 vols., 12mo, Amst. 1725). The above particulars are taken from one of the notes with which the industrious and accurate editor has enriched the work.—Vol. vi. pp. 244, 245. The English *Bayle* takes its statement from the same source.—Vol. ix., art. "Joseph Scaliger," 1739.

†" *Calvin a tres-bien faict de ne rien escrire sur l'Apocalypse.* . . . · Calvino Pontificii non maledicunt nisi quando vident præclarum ejus ingenium, et tam præclare interpretari Scripturam ut ipsi non possint ejus præstantiam assequi. O quam Calvinus bene assequitur mentem Prophetarum! nemo melius; erat summum ingenium et judicium Calvini. *Sapit quod in Apocalypsin non scripsit.*"—Scaligeriana, pp. 60, 61 (*ut supra*).

Now, we have two alternatives to offer to Sir William Hamilton regarding this reported saying of Scaliger's: either to give it up, as being taken from a book which, however curious, can have no claim to authority, after what we have seen of its history; or, to adhere to it, and then take along with it, on the same authority, certain other sayings of Scaliger regarding the Apocalypse, which are so contradictory and senseless as to destroy themselves. Thus, in the "Scaligerana Prima" he is made to say: "This I can boast of, that I understand all those predictions which are written in the Apocalypse, *that truly canonical book*, save that chapter in which '*woe*' is repeated seven times; for I know not whether the time there meant be past or future." *
Now, compare with this the following extract—a mixture of Latin and French—from the "Scaligeriana":—" A certain pastor of Montauban published a most learned commentary, in four books, on the Apocalypse, which Scaliger gave to Uitenbogard to read. The Syrian Church does not acknowledge that book, though Scaliger has a Syriac version of it, made by order of the Maronites, which the patriarch sent him. *I scarcely believe the apostle John to be the author of the Apocalypse.* The Apocalypse was written in Hebrew. . . . . . There was a minister at Castres who expounded the whole Apocalypse. . . . . Whatever was written on the Apocalypse before the last forty years is worthless. [Does not this imply that the events of that recent period

---

* "Hoc possum gloriari me nihil ignorare eorum quæ in Apocalypsi, *Canonico vere libro*, prophetice scribuntur, præter illud caput in quo '*væ*' septies repetitur; ignoro enim idne tempus præterierit, aut futurum sit."—"Scaligerana Prima," p. 13. We are obliged to take this extract from Bishop Newton (Dissert. xxiv.), as we have not been able to get hold of the original, which, our readers will bear in mind, is a different collection from the one first published, and which lies before us.

had thrown a new light on the whole subject, in the speaker's view?] *In the Apocalypse there are only two chapters which can be understood; these are very plain, nor can their meaning be disputed. Calvin was wise not to write on the Apocalypse."* To complete the *mess* in which these extracts leave Scaliger's opinions on the Apocalypse, one has only to turn to the "*Critici Sacri*" (that well-known voluminous collection of the most learned comments on the Bible), where, at Rev. xvii. 5, will be found a paragraph from Scaliger, in French, in which he is quoted as speaking of what "*the apostle*" says and does not say in that verse, as having no doubt at all about the Apocalypse being the production of the apostle John !

We may seem to have dwelt disproportionately on this small matter; but we thought it right, once for all, to put down that summary way of disposing of the Apocalypse, by reference to Scaliger's reported saying about Calvin, in which, if Sir William Hamilton stands alone, it is only because his extravagant exaggeration of it leaves him in exclusive possession. But,

5. "For which"—denunciation of the Apocalypse, etc. —"they (Calvin and Beza) were applauded by *Isaac Casaubon*."

---

* "In Apocalypsin quidam pastor Montalbanensis eruditissimum Commentarium edidit quatuor librorum, quem Scaliger dedit Utembogardo legendum. Ecclesia Syriaca hanc non agnoscit, quamvis Scaliger habeat Syriacam, que le Patriarche luy avoit envoyée, quam Maronitæ vertendum curarunt. Vix credo Joannem Apostolum autorem esse Apocalypseos. L'Apocalypse a esté escrite en Hebrew. . . . . . Il y a eu un Ministre à Castres, qui a exposé toute l'Apocalypse. . . . . . Quidquid ante quadraginta annos scriptum est in Apocalypsin, tout cela ne vaut rien. *In Apocalypsi sunt tantum duo capita quæ possunt intelligi, sunt valde aperta nec potest eorum expositio negari. Calvinus sapit quod in Apocalypsin non scripsit.*"—"Scaligeriana," pp. 10, 11.

Where has Isaac Casaubon done this? He was the contemporary, and perhaps the equal of Scaliger; spent his days in editing and illustrating the most difficult Greek classics; and, with the exception of a small Greek Testament which he published in 1587, with notes (which, however, go no farther than Acts), and "Exercitations on Sacred and Ecclesiastical Matters," in reply to Baronius' "Annals," wrote nothing theological that we are aware of. Now, in this latter work we have found the Apocalypse twice, at least, quoted and reasoned from as Scripture, in opposition to Romish doctrine; and in one of these the author, having dwelt upon Rev. xix. 15, speaks of what "*the Evangelist John immediately subjoins*" in the following verse. Until, therefore, Sir William produce his authority, we must class this along with his other strange misstatements.

6. "For which he was applauded—by our countryman, *Morus*." Alexander Morus, here referred to, was born at Castres, in Languedoc, in 1616, where his father (the Rev. Mr. More, a Scotchman), was principal of a divinity college and pastor of a Protestant church. When little more than twenty, he obtained by comparative trial the Greek chair at Geneva, over competitors all nearly one-half older than himself. Three years afterwards, he became pastor and professor of theology in room of the celebrated Spanheim, when translated to Leyden; and under him the famous F. Turretin studied for several years. But superior as were his attainments in philology in that eminently philological age, it was as a French preacher that he acquired the greatest celebrity. From Geneva he was called to a chair in Holland, where he got embroiled in the famous controversy between Milton and Salmasius, as a friend of the latter. From this he went to Paris as minister

of the Reformed Church, and died in 1670. Besides sermons and expositions, he published several works which show his high and accurate learning. The passages on which our author founds are probably two which occur in his "Calvinus," a discourse in praise of Calvin, and delivered by Morus, as Rector of the College of Geneva, in 1648. In one of these passages he is extolling Calvin as an expositor above all the fathers; and having said that he had elucidated all the books of Scripture, from the first to the last, he thus continues : " To the last, I say, nor do I except the Apocalypse, which he handled not, because by not handling it he gave the most beautiful commentary upon it " (p. 13). In the other passage he is lauding Calvin for his modesty and his moderation (*modus*). This last reigns, he says, in all his works; and it was this which continually withheld him from attempting to unravel the Apocalypse, an occupation with which those who are eager to display their ingenuity are above measure delighted (p. 49).

Now, if the reader would know what is *not* meant by these sentences, he has only to observe, first, that in his "Cause of God," or Genevese "Exercitations" on the *Authority*, the *Canon* and *Integrity*, and the *Perfection* of the Scriptures, in opposition to the views of the Romanists (1653),—where, surely, if anywhere, we should expect to find such sentiments,—not one word in the direction to which Sir William Hamilton points is to be found; next, that once and again, in his other writings, the Apocalypse is referred to just as the other books of Scripture, as the word of God; and further, that in his "Notes on Certain Passages of the New Testament" (1661), the Apocalypse comes in for its share—a small share, indeed, because, as will be seen from the extracts just given, he did not conceive himself qualified to unravel its strictly prophetical portions.

With these facts before us, what do the extracts convey beyond this,—that in Morus's day, as in ours, many had wasted their time and brains to no purpose, as he thought, in attempting to open up the mysterious portions of this book; and that Calvin's moderation showed itself, as everywhere, so particularly in this, that what he was conscious he could not succeed in, he did not attempt? In other words, both Calvin and Morus had learnt what if Sir William Hamilton had attained to he would never have penned the paragraph in question—"*Non* OMNIA *possumus.*"

7. "To say nothing of *Bodinus*, etc." We have now reached the last of Sir William's laudators of Calvin and Beza for the disparagement of the Apocalypse.

And who was this Bodinus? John Bodin was a French lawyer, who attained considerable celebrity as a literary man in the sixteenth century. He was born at Angers in 1530, and died in 1596. On his political and scientific writings the most extravagant praises have been lavished, and the severest censures pronounced; for neither his admirers nor his enemies knew any measure in their treatment of him. His dabblings in theology only revealed how ill regulated his mind was on divine things. His learning seems to have been of the kind which Festus ascribed to Paul—the abundance of it somewhat turned his head. We have before us his *Dæmonomania Magorum*, which is just a mass of learned rubbish; in the appendix to which, so far as we had patience to wade through it, he seems to maintain the reality of witchcraft on what he would call natural principles.* He wrote a book which he never dared to print, but copies of which were extensively circu-

---

* But as the *Biographie Universelle* says of his "Theatre of Universal Nature," it is a mixture of bad physics and dangerous principles.

lated, consisting of a Dialogue on Religion between seven persons *—a Deist, a Jew, a Romanist, a Lutheran, a Calvinist, and some others,—in which, says the *Biographie Universelle*, the Christians are always beaten in the argument, the advantage being given occasionally to the Jews, but mostly to the Deists. Hallam "conceives him to have acknowledged no revelation but the Jewish"; but the work just referred to, says he, "was charged by turns with being a Protestant, a Deist, a Sorcerer, a Jew, an Atheist."

What such a man thought of the Apocalypse, or of Calvin's views of it, is not worth a moment's consideration. But as Sir William's credulity is in this instance kept in countenance by Dr. Henry, in his "Life of Calvin," it may not be amiss to expose the wrong that has been done to Calvin's memory by the statement in question. Dr. Henry gives *Bayle* as his authority for the passage in question, and Sir William probably found it there too; † but wishing to see in what connection it was introduced, and whether any authority for the statement was given by Bodin in the work quoted by Bayle, we turned to the treatise itself, ("An Easy Way of Acquiring a Knowledge of History,") and found it in the introduction to a chapter on the four monarchies. This notion, says the author, of four empires, however universally it has been taken up by Biblical interpreters, is an entire mistake. No doubt Daniel seems to say as much, and it would be impious to dispute his statements. But Daniel's obscure and ambiguous language may be twisted in various ways; "and in the interpretation of the prophecies, I prefer to employ the law phrase, *non liquet* (it is not clear), rather than rashly to put other people's

---

* "*Colloquium Heptaplomerum de abditis rerum sublimium arcanis.*"
† *Dictionnaire Historique et Critique.* Art. CALVIN.

opinions upon an author in matters which I do not understand. *And highly do I approve of Calvin's no less polite than prudent speech to one who asked his opinion of the Apocalypse: 'I am quite ignorant,' said he frankly, 'what so obscure a writer means; nor is it yet agreed among the learned who and what he was.'"* *

It is strange that Dr. Henry should not only not have suspected anything wrong here, but that he should have followed it up with this remark: "Very characteristic of Calvin this, whose clear acute understanding could find nothing to work upon in this book, the mysterious phraseology of which required a prophetic insight to which he pretended not." † Bayle's observation shows more shrewdness, implying some suspicion of Bodin's statement, though insinuating in his usual way a reflection on the authority of the canonical books. "I should like to know," says he, after quoting the passage, "whether Calvin said this in any of his works, or only in conversation: I should rather suppose the latter than the former; for it would have been imprudent in a man like him to declare that it was not yet fixed among the learned what sort of man it was who wrote the Apocalypse." ‡

We have no hesitation in saying that there is not one word of truth in this story; and for the best of all reasons —namely, that in his "Institutes" Calvin repeatedly quotes the Apocalypse, and "the Apostle John" as its author,

---

* "Ac valde mihi probatur Calvini non minus urbana quam prudens oratio, qui de libro Apocalypseos sententiam rogatus, ingenue respondit, se penitus ignorare quid velit tam obscurus scriptor; qui qualisque fuerit nondum constat inter eruditos."—*Methodus ad facilem Historiarum Cognitionem,* cap. vii. p. 416. (Argent., 1699.)

† "Leben Calvin's," i., p. 348.

‡ Bayle, *Dictionnaire:* Art. CALVIN, p. 770.

without breathing a syllable of doubt either about its canonicity or its authorship. We all know what stories were trumped up by the Romanists about Luther, Melancthon, Beza, Calvin, and all the Reformers. No matter how malicious and how palpable the lie was, it served its purpose at the time; and some of them constitute the stock-intrade of some nominal Protestants, who take an inexplicable pleasure in parading whatever they can find prejudicial to their character. No doubt this is one of these Romish forgeries. Nobody can believe it except those who prefer hearsay allegations to a man's own published statements, or such as, like Dr. Henry, take it on trust, without the least inquiry, from a writer whose sceptical turn it too well suited to pass it by, and who himself was indebted for it to one of the same turn of mind.

So much for Bodinus.

8. "*Joseph Scaliger*, of the learned the most learned, rejecting also the Epistle of James, did not believe the Apocalypse to be the writing of St. John, and allowed only two chapters to be comprehensible." We have sufficiently disposed of Scaliger already, and of the two comprehensible chapters by anticipation; and having nothing to do here with the allusion to Scaliger's rejection of the Epistle of James,* we now hasten to the last of Sir William Hamilton's authorities against the Apocalypse.

---

* Here our author's information is taken, as before, from the *Scaligeriana*. If the learned talker uttered what is there ascribed to him, it does him no great credit. Luther said strong enough things about the Epistle of James, and indeed his way of testing the books of Scripture was sufficiently reprehensible, as is now universally acknowledged. But when Scaliger charged the author of this Epistle with "*great impudence*,"—if indeed he did so,—the garrulous vituperator seems to have been in one of those eminently splenetic moods which were too usual with him. The passage is as follows :—

9. "While *Dr. South*, a great Anglican authority, scrupled not to pronounce it a book (we quote from memory) that either found a man mad or left him so."

This well-known witticism of South is here quite correctly reported; but as Sir William's memory has reverted to it in the phraseology which Newton has rendered so familiar to his multitudinous readers, and in the same page of his "Dissertations" which contains the passage about Scaliger's commendation of Calvin, it is probable that our author has been originally indebted for it to the popular Bishop of Bristol, rather than the facetious Prebendary of Westminster. "A celebrated wit and divine of our own church (says Bishop Newton) hath not scrupled to assert that that book either finds a man mad or makes him so." * The remark of South to which reference is made occurs in the sermon on "The Nature and Measure of Conscience," and is as follows: "3. Because the light of natural conscience is in many things defective and dim, and the internal voice of God's Spirit is not always distinguishable, above all, let a man attend to the mind of God, uttered in his *revealed Word:* I say, his revealed Word. By which, I do not mean that mysterious, extraordinary, and of late so much

---

"Jacobi epistola plena est Judaismis; non erat recepta temporibus Eusebii." (But Eusebius himself quotes from it, and calls it "the Scripture.") "Est ab homine Catechumeno composita, quam imperitus ex omnibus aliis Scriptoribus Ecclesiasticis Canonicis collegit; et initio habet mirum de 12 tribubus; nescio quid dicam, sed magna est impudentia vocasse se Jacobum qui non est."—*Scaligeriana*, p. 167. The internal evidence here relied on against the Epistle has been abundantly refuted. A difficulty in the early church as to which of the Jameses was the author of the Epistle, and the *apparent* countenance given by it to the merit of works, were the sole cause of the early doubts about this Epistle.

* "Dissert. on the Prophecies," No. xxiv., p. 2.

studied book called the Revelation, and which, perhaps, the more it is studied the less it is understood, as generally either finding a man cracked or making him so; but I mean those other writings of the prophets and apostles, which exhibit to us a plain, sure, perfect, and intelligible rule—a rule that will neither fail nor distract those that make use of it." * On this passage, as quoted for the purposes of Sir William Hamilton, we submit the following remarks :—

First, Supposing it to express the deliberate judgment of South on the canonical claims of the Apocalypse, it is absurd to adduce it as the verdict of "a great Anglican authority." In his own line, South is inferior to none. As a preacher, his intellectual power, his clearness and force of language, his command of sarcasm and wit, were almost unrivalled. But, in reckoning up the authorities on a Biblical question, who would ever think of naming him? and, on such a point, to designate him as Sir William Hamilton has done would only be to get laughed at.

Second, We have the best of all evidence to prove that South never meant, by this witty remark of his, to cast the least reflection upon the canonical authority of the Apocalypse. *The texts of three of his published sermons are taken from the Apocalypse.* The subject of one of these discourses—and a powerful one it is—is "The Happiness of being kept from the Hour of Temptation"; and the text is Rev. iii. 10, "Because thou hast kept the word of my patience, I also will keep thee from the hour of temptation, which is coming upon all the world to try the inhabitants of the earth" (*sic*). The second paragraph of the sermon

---

* "Sermons preached upon Several Occasions." By Robert South, D.D. (Lond., 1845.) Vol. i., pp. 191, 192.

begins thus: "The occasion of the words is indeed particular, as *containing in them a prediction of the sad and calamitous estate of the Church, under the approaching reign of Trajan, the Roman emperor;* but," etc.* The subject of another sermon is, "The Lineal Descent of Jesus of Nazareth from David, by his blessed Mother, the Virgin Mary," a subject which the preacher handles with his usual acuteness; and the text of it is Rev. xxii. 16, "I am the root and the offspring of David, and the bright and morning star." The sermon begins thus: "*The words here pitched upon by me are the words of Christ now glorified in heaven.* . . . The nativity of Christ is certainly a compendium of the whole *goepel*, in that it thus both begins and *ends it*, reaching from the first chapter of St. Matthew *to this last of the Revelation; which latter, although it be confessedly a book of mysteries and a system of occult divinity, yet surely it can contain nothing more mysterious and stupendous than the mystery here wrapped up in the text,* where we have Christ declaring Himself both the root and the offspring of David."† The thing to be observed here is not the explicit testimony borne to the full canonical authority of the Apocalypse, or rather the use made of it as Scripture without an allusion to the doubts even of others—that is obvious enough; but the vindication of the Apocalypse from the objections taken to it in consequence of its mysteriousness. It *is* mysterious, says South, but not more so than the mystery of the Incarnation, on which all salvation depends. No man, speaking thus, could deliberately mean to disparage the Apocalypse. But how, then, are we to explain the witticism of which Sir William Hamilton and others make such a handle? We are far from defending it. Coming from the pulpit, and

---

* Vol. ii. p. 27 (*ut supra*).   † Vol. i., p. 507.

from one who both held it canonical Scripture and used it as such, it was in the highest degree unbecoming and reprehensible. But if the question be, How are we to *explain* it? we answer,—

Third, It is beyond all doubt one of those severe reflections upon the Commonwealth men, which South could never resist introducing into his sermons whenever he found an opportunity, or even a pretext for it. His most pungent satire, his wittiest sallies, his occasional approaches to impassioned declamation, are reserved for this theme; and if by means of some spicy anecdote of Cromwell and his preachers, he can hold up the whole party to contempt, he is in his element. Now, it is well known that among the Scriptural themes upon which the revolutionary ferment set the serious minds of all friends of the Commonwealth a-working was *prophecy,* and the Apocalyptic prophecies in particular occupied a chief place. The press teemed with works on this subject, some of them wild in the extreme; and the fierceness of the revolutionary flame was unquestionably fanned to some extent by the Apocalyptic speculations in which some ardent supporters of the antimonarchical party indulged. Of course this was fitted to inspire such a mind as South's with distaste, not to say disgust, at the whole subject. This is one of the injuries which raw and wild speculations on prophecy invariably inflict upon the prophetical portions of Scripture, and it would be well if all students of prophecy would bear it in mind. That South meant simply to express this feeling, and to do it in a way which would inspire others with the same, seems quite plain from the context. Speaking of the defectiveness and dimness of the light of conscience, and the difficulty of always distinguishing the internal voice of God's Spirit, he bids us "above all, attend to the mind of God in *His revealed Word*"; and then, repeating his words, he adds, " By this I do not mean that mysterious, extraordinary, *and of late so much studied* book called the Revelation," etc. Now, as nobody could suppose that by God's revealed Word he meant the Apocalypse, it is plain that this unexpected reference to that book is just *lugged in on purpose* to have a hit at the crack-brained Apocalyptists, as he doubtless regarded them, whom he would have left to sink into oblivion had they not been guilty of the mortal sin of disliking the arbitrary government of the Stuarts. In saying this, we neither design to justify them nor to condemn South for his political principles; but merely to show how, with South's intense political dislikes and caustic humour, the temptation to come out with one of his characteristic sallies would carry it over his sense of what was due to the pulpit, and to what he himself regarded and used as the Word of God.*

Here we close our exposure of the recklessness of Sir William Hamilton, in his paragraph on the Apocalypse. With the single exception of *Erasmus*, the weight of whose doubts has been sufficiently considered, we have seen that he is wrong in *all his authorities* against the canonicity of the Apocalypse. *Calvin* owned it; *Beza* owned it; *Scaliger*

---

* In his third Apocalyptic sermon—which we had not observed till after we had expressed, in the above paragraph, our theory about the Commonwealth-men—the author puts it beyond all doubt that he had that party exclusively in view, by repeating, with more crushing severity, but without the witticism, the very sentiment which has been caught up and reiterated in so many circles to the prejudice of the Apocalypse. The text is Rev. ii. 16, "Repent, or else I will come unto thee quickly," etc. On this text, speaking of the opinion of a learned man, that the predictions of this book were all designed to have their completion within two hundred years after their delivery, he says: " Now, if the judgment of this learned man stands, as it owned it—if we may take his own word for it above the *talk* ascribed to him; *Casaubon* owned it; " our countryman, *Morus*," owned it; even *Bodinus*—if we may judge by his quotations from it in his " Dæmonomania "—had the same faith in it as in other portions of Scripture, which perhaps was none at all; and, finally, that "great Anglican authority," *Dr. South*, owned it. Such blundering is shameful in one who professes such an intimate acquaintance with the literature of theology, and volunteers to act as a guide of the blind, a light of those that are in darkness, an instructor of the foolish, a teacher of babes. It is even ridiculous.

Here is the case. Some of the most distinguished theologians have found in the Apocalypse what Peter found in Paul's Epistles, δυσνόητα, "things hard to be understood." While entertaining not a doubt of its canonical authority, and repeatedly quoting it as Holy Scripture, they have been unable to find their way through it as a prophetic delineation of the fortunes of the Church, and have been candid enough to say so. Sir William Hamilton meets with such acknowledgments, in the course of his multifarious, but in theology superficial and inaccurate, reading. He meets with them, in hardly one instance, in their own works—for we must suppose him to be an honest man—but in the references

---

hath the countenance of reason and the express words of the text, then what must become of the bloody tenets of those desperate wretches who for these many years have been hammering of blood, confusion, and rebellion *out of this book*, from a new fancy that they have of Christ's coming? Thus ruling their lives not by precepts but prophecies, and not being able to find any warrant for their actions in the clear and express word of law or gospel, *they endeavour to shelter their villanies in the obscurities and shades of the Revelation*—a book intricate and involved, and for the most part never to be understood; and upon which, when wit and industry has done its utmost, the best comment is but conjecture." (Vol. ii., p. 304.)

made to them by other writers, some of them of the most second-rate and untrustworthy character; and these honest acknowledgments are stupidly confounded with "doubts or denials of the canonicity of the Revelation," proffered as information to an ignorant antagonist, and introduced with an expression of astonishment that he should need to be told such things—said things being, with one solitary exception, a series of bungling misstatements! Sir William's benevolent eagerness to enlighten the ignorant outruns his discretion. It did so rather notably once before, in the publication of a pamphlet at a memorable moment, full of learned illustrations, which somehow failed to convince the parties for whose special illumination it was intended. There was one formidable individual who did for Sir William then, what our slender power has been quite sufficient to do for him now—he showed him up as a theological pretender.

Nor will the Apocalypse suffer from this or a hundred such attacks. Even though the testimonies adduced against it had been genuine, as we have found them apocryphal, they are, with two exceptions, anything but formidable. We could have furnished Sir William with a more serious list. But over against these we could easily have placed an array of authorities which every competent judge would allow to be triumphantly superior, in point of weight, down to the most recent and distinguished critics in Germany. Much has the Apocalypse suffered, on the one hand, from the wild comments which have been dignified with the name of "Keys" and "Expositions," and on the other from the severity with which the most modest attempts to clear up its difficulties are by some denounced—not to speak of the advantage taken of both by the enemies of this book to hold it up to contempt. But in spite of all this, it will vindicate its own claims, and continue to shine in its own lustre; it will command increasing interest, and derive light from the march of events; its incomparable scenes, its celestial strains, its soul-stirring encouragements and appalling denunciations, even the unearthly grandeur of its language, will inspire its unsophisticated readers, though unable to thread its mysterious mazes, with courage to fight the good fight of faith and lay hold of eternal life —will nerve the hosts of the Lord for the great conflict between light and darkness, which is to issue in the rout and ruin of the phalanx of evil—will tide the Church over the last brief wave of trouble, and see it into unclouded light, unruffled repose, and everlasting glory.

THE STRUCTURE OF THE APOCALYPSE
AND ITS PRIMARY PREDICTIONS.

" The secret things belong unto the Lord our God: but those things which are revealed belong unto us and to our children for ever, that we may do all the words of this law."—DEUT. xxix. 29.

" I have yet many things to say unto you, but ye cannot bear them now. Howbeit when he, the Spirit of truth, is come, He will guide you into all the truth : . . . AND HE SHALL DECLARE UNTO YOU THE THINGS THAT ARE TO COME."—JOHN xvi. 12, 13.

THE STRUCTURE OF THE APOCALYPSE.

THE artistic structure of this book is one of its most striking features. It stands out even in its first sentence, announcing as it does Whose it is—it is "the revelation of Jesus Christ"; from Whom received—God gave it to Him; for what purpose—"to shew unto His servants the things which must shortly come to pass"; through whom communicated—"He sent and signified it by His angel" (compare xxii. 8); and to whom—"to His servant John."

After a salutation—in form quite unique, and inexpressibly grand—the writer quietly tells his readers where, how, and when he received this revelation. He was in the isle of Patmos (a little rocky island in the Ægean Sea)—banished, probably, for his fidelity to Christ. It was "the Lord's day," and "he was in the Spirit,"—his natural faculties in abeyance, and his whole inner man taken possession of by the Inspiring Spirit. In this state he heard a Voice behind him as if through a trumpet, telling him Who was speaking to him. "What thou seest (said the Voice), write in a book, and send it to the seven churches" of (Proconsular) Asia—naming them. On turning to see the Speaker, a scene was spread out before him every detail of which is here set down; the effect produced upon himself; and what his once dead but now risen and glorified Lord did and said to him, whose preciousness to every Christian heart is indescribable.

Under three heads his materials were to be arranged. "Write the things which thou sawest"—that is, this vision (i. 1—20); "and the things which are"—the existing state of the seven churches (ii., iii.); "and the things which shall be hereafter"—the strictly prophetic part of the book (iv.—xxii.).

As to the Epistles, the artistic structure of them strikes every reader. (1) Each Epistle is addressed by its glorified Head under some one of the symbols by which He is described in the opening vision, selected with express reference to the state of that particular church. (2) Each church is addressed through its "angel," or presiding minister. Yet (3) what is said to one church is expressly said to be meant for all; nay, to whoever has an ear to hear. (4) Each Epistle, though Christ's letter to that church and all the rest, is at the same time "what the Spirit saith unto the churches"; for He "takes of the things of Christ, and shews it unto us." (5) Each Epistle—whether for praise or blame —*begins* with the solemn words, "I know thy works," to summon breathless attention to what He whose "eyes are as a flame of fire" has to say to that church; and each Epistle *closes* with a *promise*, suited to what had been said of it; and this alike to those in the *best* state (as Thyatira and Philadelphia), and in the *worst* (as Laodicea). And, what is remarkable, (6) the promise to the worst as well as to the best is addressed "to him that *overcometh*," to assure all alike that, though the Christian life in every living church is a struggling life, it will prove in the end a victorious life.

But it is not till we come to the strictly prophetic part of this book that its artistic structure is seen to be, not only a feature of peculiar interest, but an indispensable key to the right understanding of the march of events. For so shifting and interpenetrating do its visions seem to be to any but the closest and most patient students, that one almost despairs (as indeed most do) of finding his way through its intricacies until the scheme of its structure begins to open upon him.

The *Choral Hymns*, which throw so great charm over this book, are designed, I believe, for the same purpose as the *Chorus* in the Tragedies of the Greek Theatre.

"The chorus (says K. O. Müller) represents the *ideal spectator*, whose mode of viewing things was to guide and control the impressions of the assembled people. The chorus . . . appears, in this kind of song, in its appropriate character, namely, to express the sentiments of a pious and well-ordered mind in beautiful and noble forms." *

Even so here: the Choral Hymns are designed, I believe it will be seen, to convey to the reader,

---

* *History of the Literature of Greece*, from the German of K. O. Müller. By Sir George Cornwell Lewis. 2nd ed., 1847, p. 311. See also Smith's "Dictionary of Greek and Roman Antiquities": Art. *Chorus*.

ITS STRUCTURE.

not only in "beautiful and noble forms," but in celestial strains, a general impression of what the symbolical visions are intended to teach.

The first of these hymns I call *The grand Inaugural Hymn*,—not so much because it extends through the first two chapters of the strictly predictive part of the book, but because it will be found to cover the whole ground of the Prophecy—embracing all that Infinite Wisdom saw it fit that the Church should know of its future fortunes. And if so, it will follow that all the successive Visions of this book are but subdivisions of this one.

The following notes on the leading features of this great hymn will make this clear. It should be noted that in its *style*, prose phraseology is studiously set aside, to make way for the stately significance of symbol.

The curtain rises majestically, disclosing a Throne, with One sitting on it in dazzling brightness—the living God, but exclusively in the character of *Creator*. Round about this Throne are "four and twenty thrones," with "elders" sitting on them, who, as they tell us themselves (v. 9), are the redeemed from among men, in two sections—those before Christ, among "the twelve tribes of Israel" (xxi. 11), and those after Christ, represented by "the twelve apostles of the Lamb" (xxi. 14).* What right they have to sit on thrones, and how they come to be "clothed in white raiment," they tell us themselves, in the song we find them singing at the outset of this book: "Unto Him that loved us, and washed us from our sins in His own blood, and hath made us unto our God *kings*." But they are "priests" as well; and that is here too, I think, though I gather this only from the function assigned to them under the symbol of the "four living creatures." For, as will be observed, all the worship celebrated in these two chapters is *led* by them: see iv. 9, 10, and v. 14.

When we pass on to chap. v. the scene changes from *Creation* to *Redemption*. In the right hand of Him that sat on the throne was seen "a *book*, written within and on the backside" (full of matter, even to overflowing), "sealed with seven seals," each bearing its own burden, but as yet a profound secret to all creation. This is set forth with wonderful sublimity, by a challenge thrown out by "a strong angel" to every creature, but accepted by none, to open or even to look upon this book; by the much weeping of the Seer at the hopelessness, to all appearance, of its contents ever being disclosed; and then by the marvellous way in which relief came to him. "Weep not," said "one of the elders"—for they were the party most interested in its contents: "behold, the Lion of the tribe of Judah, the Root of David, hath conquered" ($\dot{\epsilon}\nu\iota\kappa\eta\sigma\epsilon\nu$)

---

* Compare xv. 3, "the song of *Moses*," and "the song of the *Lamb*."

the right "to open the book, and to loose the seals thereof. And, lo, in the midst of the throne and of the four living creatures, and in the midst of the Church (as the central object), stood a Lamb, as though it had been slain (in all the freshness of its sacrificial virtue), having seven horns (omnipotence) and seven eyes (omniscience), which are the seven Spirits of God (the fulness of the Holy Ghost, in the hands of Christ, John vii. 39; xvi. 13—15) sent forth into all the earth," to conquer it for Him "whose right it is to reign."

Such was He who, in the conscious majesty of His right, "came and took the book out of the right hand of Him that sat upon the Throne." Whereupon "the four living creatures and the four-and-twenty elders (for it was *their* future that was wrapt up in that book) fell down before the Lamb, having each one a harp and golden vials,* full of odours which are the prayers † of the saints, saying, Worthy art Thou to take the book, and to loose the

---

· * There is no need, I think, to change the familiar word "vial" here into *bowl* (as in the R.V.). For though the vessel meant is more of a *bowl* shape, it is the *contents* (not the *shape* of the vessel) that is of any consequence.

† The word "prayers" here is used for all the outpourings of the heart to God, whether in prayer (strictly so called) or praise. Here it is certainly praise. The same is true of the corresponding Hebrew word in Ps. lxxii. 20, where the whole psalm is one of praise.

seals thereof: for Thou wast slain, and didst purchase (ἠγόρασας) us unto God by Thy blood, out of every tribe and tongue and people and nation, and madest us to be unto our God kings and priests:* and *we shall reign on the earth.*†

So much depends upon the sense in which we understand the last words, that to mark this I have printed them in italics. The question is, Have we here the expectant attitude of saints already in heaven, exulting in the prospect of descending at some future time to the earth, to reign on it? Or, although the *scene* is laid in heaven, have we a vision of the *Church militant* here on earth, and (at the time when this book was written) only struggling into visible existence, against determined opposition, and joyously anticipating the time when the kingdom and the dominion under the

---

* I retain here the received text and the A.V., believing that the text followed in the R.V. is a corruption of the true text. The words are taken from Exod. xix. 6, "Ye shall be unto Me a *kingdom of priests* (or with the LXX. "a royal priesthood"). If this be studied along with Rev. i. 6, and v. 10 (as given in the R.V.), it will be seen, I think, how the corruption arose.

† As to the *future* tense—"we shall reign"—I am persuaded that even the authorities that read the *present* tense, "we reign," meant it in a future *sense*.

whole heaven would be given to "the people of the Most High"—and then issuing in glory?

To me this last view gives the true conception of this song, and of the Inaugural Hymn itself, which forms the burden of the seven sealed books.

Before passing away from this Vision, the reader should note the part assigned in it to the *angels*, who, though they have no share (and need none) in the work of redemption, yet "desire to look unto these things" (1 Peter i. 11), in which they behold "the manifold wisdom of God" (Eph. iii. 10), and a love otherwise inconceivable. They stand in this vision, as is fitting, *outside* the circle of the living creatures and the elders (v. 11); and as they behold the ransomed host falling prostrate before the Lamb that was slain for them, they pay along with them their own tribute of adoration. And if there is joy in heaven over one sinner that repenteth, how much more when all arrive at their final home, to be their companions for ever! The whole scene closes with the *Amen* of the redeemed.

Come we now to the more formidable part of our task—to see whether, on the basis of the internal structure of the book itself, once solidly laid, it may not be possible to find, in outline, *the fortunes of the Church historically predicted*. And surely, if this "book of the revelations of Jesus Christ" was given for the express purpose of "shewing unto His servants the things which were shortly to come to pass," such an attempt ought not to be deemed either presumptuous or hopeless.

The first writer who threw any light upon this book to me was the learned JOSEPH MEDE (1586—1638).* 'Keys' to the Apocalypse that open nothing we have in abundance—plunging into the interpretation of the book without first settling *the limits within which alone interpretation can legitimately be looked for.* But Joseph Mede, by his *Clavis Apocalyptica,*† at once set me upon my feet. Here I found a starting-point, proceeding from which it seemed to me possible, if not to reach, yet to approach the goal.

"*The Apocalypse* (says Mede) *hath marks and signs sufficient, inserted by the Holy Spirit, whereby the Order, Synchronisms and Symbols of all the Visions may be found out and demonstrated, without supposal of any interpretation whatsoever. The order and synchronisms, thus found out and demonstrated, as it were, by* argumenta intrinseca, *is the first thing to be done, and forelaid as a foundation, ground, and only safe rule of interpretation. If the order be first fixed and settled, out of the indubitable character of the letter of the text, and afterwards interpretation guided, framed, and directed by that order, then will the variety of interpretations be drawn into very narrow compass. This is the method which I*

---

* "The Works of the Pious and Profoundly Learned JOSEPH MEDE, B.D., sometime Fellow of Christ's College, Cambridge." (Lond., folio, 1677).

† CLAVIS APOCALYPTICA *ex innatis et insitis Visionem characteribus eruta et demonstrata.*

endeavoured to represent in my Scheme [his Clavis], and demonstrate in the Tractate annexed,* in which therefore you shall find all interpretations set apart, and, as it were, discharged, and all the reasons [I assign] to be founded upon the bare letter of the text; taking no notice at all of any event or interpretation whatsoever, but leaving all at liberty; only reserved that the order and synchronism which I represent out of the text be no ways violated thereby: and so let the interpretation be whatsoever it may be." †

Here, surely, is common sense; and Mede, in his *Clavis*, working on this principle, proceeds to lay down a series of *Synchronisms* (or contemporaneous predictions), which may be proved from the text itself to be so. In this way, shutting himself in every case within in *the conditions of the question*, he refuses to look at any interpretation lying *outside* of these conditions. Thus limited, he gives us his own interpretation of the events predicted,—in which he admits he may be wrong; but leaving every one to judge for himself provided only that he keeps within the limits prescribed by the text itself.

---

* *In S. Joannis Apocalypsin Commentarius*, ad amussim *Clavis Apocalypticæ*.
† "Select Remains," p. 581.

Following, at a humble distance, this master of Apocalyptic interpretation, I now proceed to the task I have undertaken, beginning, of course, with—

THE SEVEN SEALS.

These, as I have said, cover the whole ground of prophetic disclosure in this book. From which it follows that the two Septenaries that come next in order (the *Trumpet* series and the *Vial* series) are but *subdivisions* of this all-comprehensive one.

But, besides these, there is a distinct set of Visions—not arranged in Septenaries, and having a symbolism of their own. These will be found to be of an *Explanatory* nature— designed to *fill up in detail* what in the Septenaries is presented only in dim outline. These Explanatory Visions extend from chap. x. to xiv. inclusive, after which the Septenary form is resumed in chap. xv. and xvi.

So indispensable are these Explanatory Visions to the right apprehension of the march of events, that, alongside of all I have to say on the Seals, it will be necessary to take in more or less of the Explanatory matter.

One other device of this book may be noted here. When any of the great epochal events is to occur, *it is first announced in a burst of song*, or similar form, *the details of all that lead up to it being reserved for subsequent disclosures.*

We may now proceed to the predictions of this book in order.

THE FIRST SEAL.

Happily, interpreters are pretty well agreed about the general import of the symbolism here employed. The rider on the white horse, with a bow in his hand and a crown (of *victory*—στέφανος) on his head, "going forth conquering and to conquer," is *Christ in the Gospel*, going forth to conquer, not *Judaism*, which (as we shall presently see) had already fallen, but the *Paganism* of the Roman Empire.

Passing the next three Seals (for I am not writing a commentary, but an A B C), let us see what we can get out of

THE FIFTH SEAL.

"And when He had opened the fifth seal I saw under the altar (of sacrifice) the souls of them that had been slain for the word of God, and for the testimony which they held; and they cried with a loud voice, saying, How long, O Master,* the holy and true, dost Thou not judge, and avenge our blood on them that dwell on the earth?" (vi. 9, 10).

This, surely, is not the cry of the few martyrs to the fury of a fanatical Judaism before the fall of Jerusalem. To me it speaks of a persecution so protracted and so bloody as to extort from the slain witnesses of Christ a cry of astonishment that no sign of deliverance was appearing. I hear in it the cry of whole centuries of *Pagan* persecution, well-nigh exhausting the patience and faith of the saints. But the answer to the cry is even more significant of this: "And there was given them to each one a white robe; and it was said to them, that they should rest yet a little time until their fellow-servants and their brethren, that should be killed as they were, should be fulfilled" (vi. 11).

Yes, dear souls, the voice of your blood, like

---

* Δέσποτης—"master," or "proprietor,"—of *slaves* as his purchased property. As such, it is used of Christ, whose purchased property His people are (1 Cor. vi. 20; 2 Peter ii. 1); and here as the Avenger of their wrongs (compare Rom. xii. 19).

Abel's, has cried to your Master from the ground, and it *shall* be avenged, though not yet; for another cycle of bloody persecution has to follow yours, and then will come the time of vengeance for both!

If historical prediction is to be found in this book at all, surely it is here. Church history tells at least this stern reality,—that in succession to the centuries of *Pagan*, there followed a period of *Papal* persecution, more than thrice the length of the former.

THE SIXTH SEAL.

" And I saw when he opened the sixth seal, and there was a great earthquake; and the sun became black as sackcloth of hair, and the whole moon became as blood; and the stars of the heaven fell unto the earth, as a fig tree casteth her unripe figs, when she is shaken of a great wind. And the heaven was removed as a scroll when it is rolled up; and every mountain and island were moved out of their places. And the kings of the earth, and the princes, and the chief captains, and the rich, and the strong, and every bondman and freeman, hid themselves in the caves and in the rocks of the mountains; and they say to the mountains and to the rocks, Fall on us, and hide us from the face of Him that sitteth on the throne, and from *the wrath of the Lamb:* for *the great day of their* wrath is come; and who is able to stand?*" (vi. 12—17).

The Old Testament has made us familiar with the symbolism here employed, to express a great *revolution* (Isa. xiii. 6—10, 13, 14, etc.); while the closing words, which I have italicised, tell with awful significance the nature and cause of the change. While the great red dragon was inspiring his Imperial agents to make the blood of the martyrs for centuries to flow like water, little knew they that they were kindling "*the wrath of the Lamb*,"—strange word! But they were at length made to *feel* it in their expulsion, one and all of them, from place and power in the Empire, and to see the hated thing in the throne of the Cæsars.

Turn we now to the *Explanatory* chapter—xii.—which tells the same tale in a symbolism of its own.

---

* This is evidently the true reading here.

THE WOMAN AND THE DRAGON.

"And a great sign was seen in heaven; a woman arrayed with the sun, and the moon under her feet, and upon her head a crown of twelve stars; and she was with child: and she cried out, travailing in birth, and in pain to be delivered" (xii. 1, 2). The symbolism here is easy and beautiful: it is the Church, "the bride, the Lamb's wife," in all the freshness and vigour of youth, about to give birth to a progeny of Christians, that are to rule all nations, and fill the earth.

But there was seen another sign in heaven,— "a great red dragon, having seven heads and ten horns, and upon his heads seven diadems" (διαδήματα—royal crowns*). This is the Roman Empire, in its *unbroken* supremacy, with its seven forms of government—that of Emperor being the existing one when this book was written. But, whereas it is the "great red dragon" that has the heads and the horns, this is to signify that the Empire is to be seen inspired by "that old serpent, the devil," to

---

* Not στέφανους, crowns of victory (as in vi. 2).

believe that Christianity was a religion hostile to the Empire and not to be tolerated; and so he is "red" with the blood of the Christians. Accordingly, edict after edict was issued, ordering every known Christian to be put to death. The effect of this may be imagined. Many fell away, "denying the Lord that bought them"; and as the shepherds of the flock would be the first to be seized on, traitors to Christ would be found among them; and the dragon calculated on this,—for "his tail drew the third part of the stars of heaven (the ministers of the Church, i. 20), and did cast them to the earth"—or stealthily got them to betray their trust. But there was a noble army of martyrs, who counted not their lives dear unto them, but laid them down for Him that died for them. When Pliny the Younger was Propraetor of Pontus and Bithynia, on the north-eastern shore of the Black Sea (about A.D. 110), he found that Christianity had made such progress there that it had almost emptied the temples, and that at the shambles buyers of meat which had been offered to idols could hardly be got. What, then, was this kind-hearted man to do when an edict was issued by Trajan, well-meaning

Emperor though he was, to put all Christians to death? Crowds were brought before him, charged with being Christians. What answer got he? We *were* Christians, but we gave it up long since, said some; five years ago, said one, others ten, some fifteen, and some even twenty years ago. Among the tests he put them through—such as offering incense to the gods and the Emperor's own image— there was one which never failed to show who were and were not Christians: making them "*curse Christ.*" That, it was said, no real Christian could be got to do.*

"And the dragon stood before the woman when she was about to be delivered, that when she was delivered he might devour her child," —the woman's progeny, the fresh converts to Christ. And when one bloody persecution succeeded another, with its crop of martyrs, he would seem to be doing it. But a sleepless Eye was over it, and "the child was caught up unto God, and unto His throne. And the woman fled into the wilderness, where

---

* See his Letter to Trajan, asking for instructions how to proceed, with Trajan's Answer, happily preserved among his Epistles, Nos. xcvii., xcviii. (ed. Keil, 1853).

she hath a place prepared of God, that they should feed her there a thousand two hundred and threescore (prophetic) days," or 1260 years.*

At first one is at a loss to see the connexion between the two things here announced—the ascent of "the child" to the throne, and the flight of "the woman" into the wilderness for so protracted a period. But the scene that follows—which is actually the same events as are here announced, but spread out in fuller detail, and under another symbolism—will clear the matter up.

THE WAR IN HEAVEN BETWEEN MICHAEL AND THE DRAGON.

"And there was war in heaven: Michael and his angels (see Eph. iii. 10) going forth to war with the dragon and his angels (see Eph. vi. 12); and the dragon warred, and his angels,

---

* I say "prophetic" days; for I cannot stop to discuss here the theory of *literal* as against *prophetic* days. One thing is obvious, that it is the same period as is elsewhere called "forty and two months" (xi. 2), and "a time, and times, and half a time" (xii. 14); and it will be hard to show how the things predicted of this period could take place within three literal years and a half.

and they prevailed not, neither was their place found any more in heaven" (xii. 7, 8). In the deepest sense, the war was in "the heavenly places,"—the unseen regions; in other words, between the spiritual powers of good and evil, or Christ and the great enemy of souls. But, as a matter of fact, it was fought down here, in human flesh and blood, as the martyrs to dragon-inspired Paganism knew to their cost. But the result was glorious, and is told in these thrilling words: "And they overcame him (the dragon) by* the blood of the Lamb, and by the word of their testimony; and they loved not their lives even unto death" (xii. 11).

"And the great dragon was cast out, the old serpent, called the devil and Satan, the deceiver of the whole world; he was cast out, and his angels were cast down with him" (xii. 9). In other words: By the accession at length of a Christian to the throne of the Empire, Paganism was overthrown. All the diabolical influence

---

* I retain here the A.V.; for, though the preposition (διὰ with acc.) properly means "by reason of" rather than "by" or "through," the senses (as *Winer* says) are apt to run into each other,—as here, I think.

with which the dragon inspired his Pagan agents immediately ceased. No wonder that all heaven rang with joy, saying, " Now is come salvation, and the kingdom of our God, and the authority of His Christ: for *the accuser of our brethren* is cast down, who *accuseth them before our God day and night*."*

This closes the first period of predicted Church history, in symbol,—the seating of Christianity on the throne of the Cæsars in the fourth century. But the dragon's resources are far from being exhausted. Since the Christians have got the better of him, he will become a Christian himself, to betray Christ with a kiss! Was there a Judas among the Twelve, a " son of perdition " (John xvii. 12), who " did eat of Christ's bread, but lifted up his heel against Him " (Ps. xli. 9; John xiii. 18)? The dragon

---

* In the words which I have italicised the reader should note the glimpse which they give of what passes in the unseen regions of the spiritual powers of good and evil—" the heavenly places " —one of many too often overlooked or explained away. The second chapter of Job gives one example of what is here said to be going on incessantly. Usually this is viewed as but a dramatic representation of the struggle that goes on in the human breast between good and evil. But, in the light of the words I have italicised, perhaps it will be seen to be a stern reality. See Zech. iii. 1—6, and Jude 9.

ITS PRIMARY PREDICTIONS.

will know how to work even Christianity into an engine of his own against Christ Himself, and will create a "son of perdition" in the heart of the Church! (2 Thess. ii. 3.)

On the accession of Constantine, persecution of course ceased. And just as when Saul of Tarsus, from being a bloody persecutor, became a preacher of the faith which once he destroyed, one might have said, after this triumph of the Gospel: "Then had the churches peace; and, walking in the fear of the Lord and in the comfort of the Holy Ghost, were multiplied." Yes, multiplied they were: for the persecuted soon became the popular cause. Numbers flocked into the Church; but the quality was not as the quantity, for living Christianity rapidly declined. The Emperor naturally paid court to the bishops, and this proved too much for them. Clerical ambition set in, and new ecclesiastical orders were invented to correspond with the grades of office in the State—from bishops to archbishops, metropolitans, patriarchs, primates; and what that would culminate in no one needs to be told. To be Bishop of Rome in persecuting times was what no one would covet who was not prepared to be the first victim : but now it was the most coveted of all posts, at the seat of power, and ensuring the ear of the Emperor. How the gradual weakening of the Eastern Empire tended naturally to strengthen the West, and how its clergy were not slow to improve their advantage, until at length it issued in a full-blown Papacy at Rome, I leave Church history to tell.

We are now to see what is the dragon's new policy. It is twofold. In the same verse that calls upon the heavens to rejoice because of the triumph of Christianity over Paganism, we have a "woe for the *earth* and for the *sea*"; for there, it seems, we shall find "the devil" at work :—" Therefore rejoice, O heavens, and ye that dwell in them. Woe for the earth and for the sea: because the devil is gone down unto you, having great wrath, *knowing that he hath but a short time*" * (xii. 12).

---

* Here is another of those glimpses into the Unseen which should not be overlooked. The great enemy of Christ has been made to know, it seems, the limit of his reign. (Compare Matt. viii. 29: "What have we to do with Thee, Jesus, Thou Son of God? art Thou come to destroy us *before the time*?") And this shows, too, that they know their Judge to be Jesus, the Son of God, and the "Seed of the woman."

ITS PRIMARY PREDICTIONS.

["And when the dragon saw that he was cast down to the earth, he persecuted the woman which brought forth the man child. And there were given to the woman the two wings of the great eagle, that she might fly into the wilderness unto her place, where she is nourished for a (prophetic) time, and times, and half a time (=1260 years), from the face of the serpent" (xii. 13, 14).]

I have bracketed these two verses, as they are evidently parenthetical, referring the reader back to verse 6. But before going further we must go back to chap. vii., where we shall find —not the opening of the *Seventh Seal*, which we should expect, after the fall of *Paganism* (vi. 12—17), but great *preparations for the opening of it;* ushering us, in fact, into the events of the 1260 years.

PREPARATIONS FOR THE OPENING OF THE SEVENTH SEAL.

Chap. vii.

After reading the disclosures of the Sixth Seal (vi. 12—17), we expect to read: "And when He had opened the Seventh Seal, I saw"—such and such things. But, instead of this, four angels are seen holding the four winds of the earth, to keep them from bursting forth in a desolating storm, which, but for this, might destroy indiscriminately the righteous with the wicked. To prevent this an angel is seen "having the seal of the living God," to "seal the servants of our God on their foreheads"—as "living epistles of Christ, known and read of all men" (2 Cor. iii. 3).

The number of the sealed is then symbolically expressed by the square of $12 = 144,000$; for "the Lord knoweth them that are His" (2 Tim. ii. 19). They are of every tribe of the children of Israel.

I should have thought it unnecessary to prove that *literal* Israelites are not meant here. But as the whole sense of the prophecy depends on it, and even good interpreters go far astray here, I will now prove it. (1) The tribal division is quite different from that of the natural Israel. (2) Of *unconverted* Israelites it could not be said that they were "the servants of our God," and of *converted* ones there is no conceivable reason why *they* should be singled out from among all other Christians for preservation. But above all, (3) when we turn to chap. xiv.—where we find the same company of 144,000 "sealed" ones—not only are they not called Israelites at all, but they are described by such marked features of character as make it evident that they are Christ's *faithful witnesses in a period of almost universal degeneracy and persecution unto death.* *First*, they "stood with the Lamb* on the mount Zion" (where the true Israel worship), xiv. 1; and they "had their Father's name written on their foreheads," as in vii. 5. *Next*, they were virgin souls, uncontaminated by the reigning defection, "following the Lamb whithersoever He goeth" (xiv. 4); "in their mouth was found no lie,† and they were without blemish" (xiv. 5). *Again*, as such, they were not suffered to live: for they were "purchased ($\dot{\eta}\gamma o\rho a\sigma\mu\acute{e}\nu o\iota$) from the earth" (xiv. 3); or (as in verse 4) "were purchased ($\dot{\eta}\gamma o\rho\acute{a}\sigma\theta\eta\sigma a\nu$) from among men, being the first-fruits unto God and the Lamb." *Once more*, they had a song to sing, which none knew but

---

* Not "a lamb"—according to the A.V. and a heading of the text, which has next to no support.

† Not "guile"—on next to no evidence.

themselves; being a class of martyrs by themselves—martyrs not for Christianity against *Paganism*, but for Christ against *Antichrist*: in other words, they were that *second* class of martyrs which the *first* class were expressly told they must wait for, ere their own blood could be "avenged." They must "wait yet for a little time, until their fellow-servants and their brethren, that should be *killed as they were*, should be fulfilled" (vi. 11).

Well, now that *they* also are killed, what follows? Turn back to the "sealing" of the 144,000 in chap. vii., and immediately after that we read: "After these things I saw, and behold, a great multitude, which no man could number, out of every tribe, and people, and tongue, and nation, standing before the throne and before the Lamb, arrayed in white robes, and palms in their hands; and shouting, Salvation to Him that sitteth on the throne, and to the Lamb"—and all that follows to the end of that chapter. *It is this noble army of martyrs, of both classes, that is specially in view.* The language, it is true, is so catholic, that when read, with no reference to the *place* and the *connexion* in which it stands, it reads simply as a matchless description of the heavenly state, with no reference to one class more than another. As such it is read and will be read as long as there are those who have friends already there, or are themselves on their way thither. But, all the same, one has only, after reading the preceding part of chap. vii. up to verse 8, and then what follows, to observe the emphasis in verse 14—"These are they which come out of *the great tribulation*, and they washed their robes, and made them white in the blood of the Lamb"—and they will come, I believe, to see that *one class* of the glorified is specially in view, namely, the 144,000 "sealed" ones, described in chap. xiv., as "redeemed from the earth." The language, as I said, is too catholic to be *restricted* to any one class of saints; but this is a characteristic of Scripture language—that what is *specially* meant for one class is so expressed as to admit of being extended *in principle* to all of the same class.*

---

* Take, for example, the very chapter just commented on (xiv. 13): "I heard a Voice from heaven, saying, Write, Blessed are the dead which die in the Lord from henceforth: yea, saith the Spirit, that they may rest from their labours; for their works follow with them." Read simply as expressing the blessedness of

Preparation being thus made for the protection of the "sealed" against the desolating storm, whose approach is announced in vii. 1, the opening of the Seventh Seal is at length announced.

THE SEVENTH SEAL.

Chap. viii., ix.

But, instead of the usual words "And when he had opened the Seventh Seal," its *contents* were such and such, we read, "there followed a silence in heaven about the space of half an hour,"—in other words, there ensued a brief pause before the curtain rose, so to speak, for a new series of disclosures. These, we find, are seven in number, or seven great events announced by the sounding of Seven Trumpets, in the hands of Seven Angels.

THE SEVEN TRUMPETS.

It is no part of the plan of this A B C to take up these in detail. I hasten, instead, to

---

all who "sleep in Jesus," this has been the theme of funeral sermons; and most suitably. But read *in connexion with the context*, both before and after, it will be seen to refer specially to *that class* which is the burden of the whole chapter.

what will open up the whole subject of the Prophecy, in a form and with a detail not to be had in the Septenaries. I refer to the *Explanatory Visions*, the character of which I have already explained (see p. 78). These Visions begin with chap. x., and extend through several successive chapters.

Chap. x. opens with "a mighty angel" seen coming down from heaven, with characteristics which seem to identify him with Him whom the Seer first beheld, in the vision of chap. i. He bestrode "the earth and the sea," to signify his purpose to take possession of them as His own. He "had in his hand a Little Book *open*" —for its contents had already been *disclosed*, but not yet sufficiently *explained*. So he cries with a lion-roar. Whereupon "seven thunders uttered their voices." And they were *articulate* voices, for the Seer was about to write *what* they uttered. But this he was forbidden to do— not because their contents were not to be disclosed at all, but because the time for it would not be until the iniquities of the antichristian Power were full. "But in the days of the voice of the seventh angel, when he is about to sound" (his trumpet), it would then be seen that the "seven thunders" were the Seven Vials of the wrath of God upon the antichristian Power. Nor was this to be long delayed. For the angel that talked with the Seer is heard swearing by Him that liveth for ever and ever that "there should be *time no longer*" given for delay, than till "the days of the voice of the seventh angel, when he is about to sound, the mystery of God should (then) be finished, according to the glad tidings which He declared to His servants the prophets."

The Seer is now told to go and take the Little Book out of the angel's hand; which having done, he is told to "eat it up," and he would find it "bitter in his belly" (its contents, when digested, would be found very doleful), but "in his mouth sweet as honey" (as all God's words are to His own): compare Ps. xix. 10; Jer. xv. 11. This he did, and found even as said to him; and the meaning of the act is made plain to him as follows: "Thou must prophesy again"—or utter a fresh series of predictions—"over (or concerning) many peoples, and nations, and tongues, and kings"; in other words, concerning the States of Christendom.

But, since the Little Book was *open*, these new predictions could be nothing more than a fuller or more expanded representation of what had been summarily announced before —in the Explanatory Visions. The first one, in chap. xi., I reserve until the second one— xii. 13, 14, which covers more ground—has been opened up.

THE DRAGON'S NEW POLICY.

Chap. xii. 13, 14 (see p. 90).

"And when the dragon saw that he was cast down to the earth, he persecuted the woman which brought forth the man child. And there were given to the woman the two wings of the great eagle, that she might fly into the wilderness unto her place, where she is nourished for a (prophetic) time, and times, and half a time" (1260 years).

"The dragon," it will be remembered, has "the seven heads and ten horns," working through the Imperial Power of Rome, under which Christianity was born. But now that Christianity is on the throne, in the person of a

Christian Emperor, he will direct his attacks upon two quarters—the *earth* and the *sea*; for each a "woe" of pity is pronounced, for the evils that would come upon the quarters thus marked out.

"WOE FOR THE EARTH."

CHAP. xii. 12.

" And the serpent cast out of his mouth after the woman water as a river, that he might cause her to be carried away by the stream. And the earth helped the woman, and the earth opened her mouth, and swallowed up the river which the serpent cast out of his mouth" (xii. 15, 16).

What I have to say on these verses may be passed over in the meantime; for though, as I understand it, it is a necessary link in the chain of events, I am anxious to hasten on to the "woe for the *sea*," which will be found to be the burden of the whole sequel of this Prophecy, up to chap. xx.

Mede and other interpreters go astray here, as I judge. It is true that *water*, as a cleansing element, is a symbol in Scripture of divine truth (Ps. xlvi. 4; Isa. xii. 3). And as water from the serpent's mouth must needs be deadly, they see here a prediction of the deadly *Arian* heresy, which at one time threatened to strip the Faith of all that is vital in it, but which in the end fell more signally than it rose into favour; while the orthodox faith acquired even more than its former supremacy both in the councils of the Church and the support of the throne: the earth thus swallowing up the flood which the serpent cast out of his mouth after the woman. But there are two fatal objections to this. (1) The "earth" in this chapter is neither the Church nor the Christian state. (2) The "water' here referred to is explained in xvii. 15 to mean *peoples* or *nations*. Now, the "peoples" here referred to (as we shall see later on) are those who at a later stage, when Christianised, are found supporting the "harlot." At this stage they were outside of Christianity.

As I take it, they were those fierce northern races who gradually migrated southward and westward from the sterile region of Tartary, till, lying along the rich northern provinces of the empire, they by little and little fought their way into the heart of the empire, and at length broke it up. But the victors had intelligence as well as bravery, and embraced both the civilisation and the religion of the vanquished; and the fragments of the broken empire became consolidated into distinct and independent Christian sovereignties— thus preparing the way for a supremacy over them all even more real than that of the Cæsars, but entirely different. Thus did "the earth help the woman, swallowing up the flood by which the serpent thought to carry her away,"—pouring heathen races into the empire, hoping thereby to heathenise it. But again did Christianity supplant Paganism—yet here, too, as we shall see, only to be made a still deadlier enemy of Christ.

"WOE FOR THE SEA."

"And he * stood upon the sand of the sea; and I saw a (wild) beast (θηρίον) come up out of the sea, having ten horns and seven heads,† and on his horns ten diadems (royal crowns), ‡ and on his heads names of blasphemy. And the beast which I saw was like a leopard, and his feet were as the feet of a bear, and his mouth was as the mouth of a lion; and the dragon gave him his power and his throne, and great authority. And I saw one of his heads as though it had been smitten to death (Greek "slain"); and his death-stroke was healed: and the whole earth wondered after the beast. And they worshipped the dragon, because he gave his authority unto the beast. And they worshipped the beast, saying, Who

---

* He, the dragon, stood—not the Seer, according to the received text and the A.V. By this mistaken reading the reader fails to see why, in order to bring up this terrible beast, the dragon had to go to the sea for him, and is here found standing by it.

† This significant change in the order of the words is undoubtedly the genuine text.

‡ See on chap. xii. 3, p. 83.

is like unto the beast? and who is able to make war with him? And there was given to him a mouth speaking great things and blasphemies; and there was given to him authority to continue forty and two months. And he opened his mouth in blasphemies against God, to blaspheme His name, and His tabernacle, and them that dwell in heaven. And it was given to him to make war with the saints, and to overcome them. And there was given to him authority over every tribe, and people, and tongue, and nation. And all that dwell on the earth shall worship him, whose names have not been written from the foundation of the world in the book of life of the Lamb that hath been slain" (xiii. 1—9).*

In these nine verses the new power called "the (wild) beast" is described with a precision as well as fulness of detail, as if to prevent mistake either as to its *nature* or its *seat*. It will not do, therefore, to generalise it away by calling it simply the great *world-power*,

---

* That this arrangement of the words gives the true sense of the clause will be seen by comparing it with xvii. 8, where the thing which took place "from the foundation of the world" is not the death of the Lamb (for that is not mentioned at all), but the writing of the names.

which in every age is found antagonistic to the kingdom of God. It is contrary to all the principles of strict interpretation to dispose of details so specific, so varied, and so peculiar, in this way. Let us try, then, how far by strict exegesis we can approach the solution of this great question.

CHARACTERISTICS OF THE POWER CALLED THE "BEAST."

1. Whereas it is said to come up "out of the sea," this is to direct the reader back to Daniel's Vision of the "four beasts" (Dan. vii.). There the four winds of the heavens are seen breaking forth upon the great sea, and out of this convulsion "four great beasts come from the sea." These are the four great universal empires which had their origin in political convulsions; and of these the *fourth* one is the beast of Rev. xiii.

2. Whereas in xii. 3 the dragon with seven heads and ten horns has the diadems on the *heads*, but now the same beast has the diadems on the *horns*, this is to signify that the Roman Empire had, by the time there referred to, been broken up, and that its fragments had

become so many distinct and independent sovereignties.

3. Though independent of each other, these States all owned subjection to a common head, here called the beast, whose horns they were.

4. If the empire had fallen ere the diadems were found upon the *horns*, it follows that the beast of Rev. xiii. cannot be any of the emperors.

But of all the emperors that have been fixed upon as "the beast of the Apocalypse," the one who just now is the favourite with distinguished critics has the least, I think, to recommend it—the Emperor *Nero*. The critics I refer to, recognising in this book nothing strictly predictive, and seeing in it only the passing events of the writer's own time, seize upon a story then current, that Nero, instead of having killed himself—which he was known to have done—was only hiding himself in the East, and would return to reign again. This, they are confident, is the true explanation of the death-stroke which one of the heads of the beast received, but which was healed (xiii. 3). It is also the solution of the riddle, in xvii. 11, about "the beast that was and is not, and is himself an eighth, and is of the seven." "In fact" (says Professor Harnack) "this book, long thought to be the most obscure and difficult document of early Christianity, . . . is neither obscure nor mysterious. . . . Without being paradoxical, we may affirm that the Apocalypse is the most intelligible book in the New Testament ; because its author had not the individuality and originality of Paul or the author of the fourth Gospel." Indeed (adds Harnack) fifty years ago, certain scholars (whom he names) simultaneously discovered "the number of the beast—666" (Rev. xiii. 18), in the numerical value of the Hebrew letters כסר נרין (*Cæsar Nerōn*, Emperor Nero).* No doubt of it; but he might with as much truth—and as little pertinence—have told his readers that, besides *Irenæus* (end of second century), whose conjecture as to the number of the beast is as respectable as any, others from time to time have—by a like manipulation of Hebrew or Greek words—found in other words the number of the beast. I myself possess an old edition of Irenæus ("Adv. Hær."), with

---

* *Encycl. Brit.*, 9th ed. : Art. "Revelations."

copious foot-notes by the editor (*Feuardentius*), in one of which that pious Jesuit finds the number of the beast in *Luther*—as he spells the word! One might protest against the arbitrariness of searching for the number 666 in the numerical value of *Hebrew* letters in a book written in Greek. Also, since Nero did *not* return, and, consequently, the deadly wound was *not* healed, do they mean to tell us that the writer of the Apocalypse was a false prophet? But such questions would not trouble the critics of this school; for they can sit loose to such niceties.

I have spent more time on this speculation than it deserves. But if it teaches this lesson, that mere scholarship—especially anti-super naturalist scholarship—is poor furniture for the interpretation of Scripture, and least of all for such a book as this, it will not have been quite useless.

5. This new power, called "the beast," is a purely *ecclesiastical* power, and in *character* and *purpose* is a *diabolical* power: for "the dragon gave him his power, and his throne, and great authority." And "he opened his mouth in blasphemies against God, to blaspheme

His name, and His tabernacle (His Church), and them that dwell in heaven" (? saints and angels). "And it was given to him to make war with the saints, and to overcome them: and there was given to him authority over every tribe and people and tongue and nation,"—that is, over all the territories of the old empire, now Christianised (xiii. 2, 5—7).

In Daniel's vision this feature of his "fourth beast" comes strikingly out. It was "*diverse from all the beasts that were before it*"—which can only mean that it was not a *secular* or worldly power; while its blasphemies stand out in what is said of the "little horn" which "came up among" the ten horns of this beast: it "had a mouth speaking great things" (compare Rev. xiii. 5: "And there was given to him a mouth speaking great things and blasphemies"). And "because of the voice of the great words which the horn spake, I beheld even till the beast was slain, and his body destroyed, and he was given to be burned with fire" (Dan. vii. 8, 11).

6. Whereas the other three beasts in Daniel's vision were like the three fiercest of the beasts of prey, and there was no fiercer to liken the

fourth one to, our Apocalyptic beast combines in himself the properties of the other three (xiii. 2), as being the most dreadful of all; accordingly, Daniel says he was " dreadful, and terrible, and strong exceedingly " (vii. 7).

7. We have seen that all the States that succeeded to the territories of the old empire —though independent of each other—owned a common authority. And this authority, being in its own nature purely *ecclesiastical*, and so not conflicting with theirs, could be willingly recognised by all. So far, in fact, was it from being an enforced subjection, yielded in spite of its blasphemies, it was yielded to expressly in that character. For it is said : " They worshipped the dragon, *because* he gave authority to the beast ; and they worshipped the beast, saying, Who is like unto the beast? and who is able to make war with him ? " (xiii. 4).

Now, mark this terrible *compound of Church and State*; for it explains all the seeming contradictions in what is said of " the beast " in the succeeding chapters of this book. Thus, he is said to do things which *in his own sphere* he could not do. He " makes war with the saints, and overcomes them." But not a hair of their heads could he touch as an ecclesiastical power. *His* business was to summon before his proper tribunal all who dared to dispute his monstrous claims and protest against his blasphemies; and having convicted them of "heresy," to hand them over to the civil "authorities" of the states to whose jurisdiction they belonged, to be by them put to death. Thus neatly was the business managed. Most effectually was the "deadly wound" healed; the loss of *Imperial* authority in the person of the last emperor being more than supplied by *Papal* supremacy. It was a mutual gain. It gave the independent Sovereignties of the fallen empire a *unity* they much needed—a unity, too, of such strength as the empire itself never possessed. In return for this, they one and all, as Christian states, rendered implicit obedience to their ecclesiastical head. Nor was he, on his part, slow to realise his position and assert his supremacy. It was truly the master-stroke of the dragon's policy. And its development, with all its bloody results, will come up before us in the succeeding chapters.

8. The *duration* of this bestial power is the only other feature of it as depicted in xiii. 1—5

which remains to be noticed. "There was given to him authority to continue forty and two (prophetic) months"—the same period as "one thousand two hundred and threescore days" (xii. 6), and "a time, and times, and half a time" (xii. 14), or 1260 years. What this precise period is in history I inquire not here. But two things about it, which stand out on the face of it, I do notice. (1) A power whose duration is so protracted can be no single individual, and therefore must be sought in *the successive occupants of the seat of office.* (2) At whatever period we date the *beginning* and *end* of this power, no one acquainted with Church history can fail to find where the authority rested which for nearly a thousand years had a tribunal erected, before which were summoned all within Christendom who dared to dispute its claims and protest against its blasphemies, there to be tried for "heresy," and when convicted—which they always were—handed over to the civil authorities to whose jurisdiction they belonged, to be by them put to death, until the number of those who suffered martyrdom, in every form of cruelty, exceeded all computation. *Where* could be found the seat of a power from which could issue mandates so diabolical, and under the sanction of our holy religion? Church history has but one answer to this question: The city of the seven hills, "that great city which reigneth over the kings of the earth" (xvii. 18)—"the eternal city."

Having seen, at some length, how the Explanatory Vision of chap. xii.—xiii. confirms by its details all that we find predicted in the first, fifth, and sixth seals, let us now turn to the Explanatory Vision of chap. xi.

MEASURING THE TEMPLE, THE ALTAR, AND THE WORSHIPPERS.

"And there was given me a reed like unto a rod: and one * said, Rise, and measure the temple"—or "sanctuary" ($ναός$)—"the altar, and them that worship therein. But the court which is without the temple leave out without, and measure it not; for it is given to the nations," or "Gentiles" (xi. 2).

---

* The words "the angel stood" (in the received text and A.V.) are not found in the best authorities.

In Zech. ii. 1, etc., we have a similar Vision, from which we learn that "measuring the temple" is a symbol for *appropriating to the Lord* what is thus "measured off." Consequently, "the court which is without the temple, which was to be left out, and not measured, because it was given to the nations," means a worship and worshippers that the Lord would not own as His. So there is an *inner* and an *outer* circle, *a Church within the Church*—" a holy priesthood, offering spiritual sacrifices acceptable to God through Jesus Christ" (1 Peter ii. 5); and a body of hypocritical worshippers who cry, "The temple of the Lord, the temple of the Lord, the temple of the Lord are these" (Jer. vii. 4); but to whom the Lord of the temple says: "To what purpose is the multitude of your sacrifices unto Me? . . . When ye come to appear before Me, who hath required this at your hand, to tread My courts? Bring no more vain oblations: . . I am weary to bear them" (Isa. i. 11, etc.).

All this is expressed in terms of the literal temple. After the dispersion of the Jews, their religion made considerable way among the heathen, and a large number of professed worshippers of Jehovah had to be provided for in the services of the temple; accordingly, the court which was without the temple was appropriated to them. But many of these, and especially their children and descendants, greatly degenerated; and not being children of Abraham, the Jews looked down upon them very much as Christians are apt to do upon a "converted Jew."

Such is that "outer court" of Christians, called the "Gentiles." But they are worse than merely nominal Christians: for "the holy city shall they tread under foot forty and two months" (xi. 2).

This last figure is taken from our Lord's prediction, that "Jerusalem shall be trodden down of the Gentiles, until the times of the Gentiles shall be fulfilled" (Luke xxi. 24); while the period of this "treading" (forty-two prophetic months) brings us in line with all that we have found of that blasphemous Power which "makes war with the saints and overcomes them," and to whom power is given to continue forty and two months.

THE TWO WITNESSES.

"And I will give unto My two witnesses, and they shall prophesy a thousand two hundred and threescore days, clothed in sackcloth" (xi. 3).

Under the Jewish law two witnesses were required to make a valid testimony to any fact. Three would be better, but less than two would not do (Deut. xvii. 6). So that there would be left to Christ, during that cloudy and dark day, barely a valid testimony—what He Himself affectingly calls "My two witnesses"—so effectually would "the Gentiles," or Gentilised *Church*, tread under foot "the holy city." No wonder that these two witnesses had to utter their testimony "clothed in sackcloth"! Bishop Wordsworth and others take the two witnesses to mean the Old Testament and the New. But, besides that this way of dividing the Scriptures is quite a modern one, the Scriptures can with no propriety be said to prophesy clothed in sackcloth It is living witnesses for Christ that we are to see here; and what follows confirms this.

"These are the two olive trees and the two

candlesticks, standing before the Lord* of the earth" (xi. 4). To understand this we must turn to Zech. iv., with which the reader is supposed to be familiar, just as in xiii. 1 he is supposed to be familiar with Dan. vii. In Zechariah's Vision one candlestick only is seen: here they are two, because the testimony of each of the two witnesses is needed to be a valid testimony for Christ, as we have seen. In Zechariah's Vision we find that the two olive trees supply the oil that gives and keeps in the light of both candlesticks. And we have the interpretation of this from the angel that talked with the prophet: "Not by might, nor by power, but by My Spirit, saith the Lord of hosts. Who art thou, O great mountain? before Zerubbabel thou shalt become a plain: and He shall bring forth the head stone with shoutings of Grace, grace unto it" (Zech. iv. 6, 7). Yes, it is only by the indwelling of the Spirit of Christ in the few scattered witnesses for the truth, that they can hold out against that overwhelming Power which they have to withstand during that dismal period.

"And if any one seek to hurt them, fire

---

* So read the best authorities. See Zech. iv. 14.

proceedeth out of their mouth and devoureth them : and if any one shall seek to hurt them, in this manner must he be killed" (xi. 5). This is explained by what is said of the Old Testament prophets, as Jer. i. 10—"Behold, I have put My word in thy mouth: see, I have this day set thee over the nations and over the kingdoms, to pluck up and to break down, and to destroy and to overthrow, to build and to plant"; also v. 14—"Behold, I will make My words in thy mouth fire, and this people wood, and it shall devour them"; and Hos. vi. 3— "Therefore have I hewed them by the words of My mouth." In other words, what the prophets were made to utter, God would cause to be executed. Accordingly, when Christ's witnesses "prophesy"—all out of sight in the wilderness, and in sackcloth—against the vast antichristianism that stifles the Gospel, their voice is Christ's voice, and their denunciations of its blasphemies and its cruelties, their word, will certainly take effect.

"These have the power to shut heaven, that it rain not in the days of their prophecy: and they have power over the waters to turn them to blood, and to smite the earth with every plague, as often as they shall desire" (xi. 6) We have here a twofold prediction—*negative*, no rain; *positive*, to poison the waters. In the former case the reference is to the *dearth* which Elijah's prediction brought upon the land of Israel in the reign of Ahab, who did his utmost to corrupt the worship of Jehovah, as this antichristian Power would do during *his* reign. In the latter case, their turning the waters into blood, the allusion is to the first of the plagues of Egypt, and the deadly effects of it. The meaning is, that while there would be a dearth of all food for hungry souls, the teaching given forth during the reign of Antichrist would be ruinous to his subjects.

For important additions to this (omitted by mistake), see Addendum I.

MARTYRDOM, RESURRECTION, AND ASCENSION OF THE WITNESSES.

"And when they shall have finished their testimony, the beast that cometh up from the abyss (Gr. 'bottomless depth') shall make war with them, and overcome them, and kill them" (xi. 7). In xiii. 7 this is given as one of the characteristics of "the beast" (see pp. 107, 108);

but here it is that one effort by which he hopes to finish the business of putting an end to all open witness for Christ.

"And their dead bodies shall lie in the street of the great city, which spiritually is called Sodom and Egypt, where also their* Lord was crucified" (xi. 8).

For a dead body to lie unburied in the street of a great city, exposed to the gaze of all passers-by, is a mark of supreme contempt in every country. That no literal city is here meant, it is needless to say; and the three characteristics of it are so striking that it would seem impossible to doubt what it is. Spiritually called, it is "*Sodom*," for its filthiness; spiritually, it is "*Egypt*," for the oppressions of God's people: "where also their Lord was crucified,"—for in the slaughter of these His faithful witnesses they "crucify the Son of God afresh, and put Him to an open shame" (Heb. vi. 6). Where are these three characteristics to be found but in *Roman Christendom*?

"And from among the peoples and tribes

---

* Such is the genuine reading here.

and tongues and nations men look upon their dead bodies three days and a half (symbolically understood), and suffer not their dead bodies to be laid in a tomb. And they that dwell on the earth rejoice over them, and make merry: and they shall send gifts one to another, because these two prophets tormented them that dwell on the earth" (xi. 9, 10).

This is simply a hyperbolical way of expressing the exulting confidence of the antichristian worshippers of the beast, that now at length they have got rid of those "tormenting" heretics along the whole line.

"And after the three days and a half the spirit of life from God entered into them, and they stood upon their feet; and great fear fell upon them which saw them. And they heard a great Voice from heaven saying unto them, Come up hither. And they went up into heaven in a cloud, and their enemies beheld them" (xi. 11, 12).

As the slaying of the witnesses symbolised *the extinction of their testimony*, so their resurrection and triumphant ascension into heaven, in the sight and to the terror of their enemies, means the resuscitation of their testimony to Christ and His saving truth, in the unexpected appearance of a body of living witnesses—no longer "clothed in sackcloth," but boldly proclaiming buried truths.

"And there was a great earthquake, and the tenth part of the city fell; and there were killed in the earthquake seven thousand persons (Gr. 'names of men'): and the rest were affrighted, and gave glory to the God of heaven" (xi. 13).

Since "the great city" of verse 8 was not Rome itself, but Roman Christendom, the fall of the tenth part of it can mean nothing else than the revolt of a vast number from its authority—their secession from its pale; while the "earthquake" that caused it means its convulsive character. It reminds one of Ezekiel's Vision of the dry bones—"very many, and very dry in the open valley." Could such bones live? Yes, when they were "prophesied to." Then the bones "came together"; and the prophet having prophesied to the wind, breath came into them, and they lived and stood up a great army. This meant the whole house of Israel resuscitated from their dead state of idolatrous departure from Jehovah:

this is the whole testimony of Jesus Christ, slain in the persons of Christ's "two witnesses," or all that openly remained for Him, but now resuscitated in a vast body of living witnesses, not afraid of the threats of Antichrist. Where, now, is his proud boast of the "marks of a true Church," possessed solely by him? *Visibility*, we have seen, is exactly what was *wanting* in "the temple, the altar, and they that worship therein"—the only Church which Christ "measured" off as His own, during the predicted period of "forty and two (prophetic) months" (xi. 1, 2). This was *no mark* of a true Church. And now its other boast of *unity*, that also is gone. The "earthquake" revealed the gap made in the hitherto unbroken ranks of the antichristian Church, breaking off from Rome just the most intelligent and powerful countries of Europe—Germany and Switzerland, the United Netherlands and Great Britain. And what is of special significance is, that *this breach has never been healed*; nor have the three succeeding centuries given any indication of a *retrograde* tendency, but much the reverse. There have from time to time been ebbs and flows of the tide; but eyes that can read the inner springs of human actions see clearly whither the light and life and health and strength of Christianity are resistlessly tending, under the guiding Eye of Him who gave it birth.

In fact, the sequel of this chapter speaks as much: "The second Woe is past: behold, the third Woe cometh quickly. And the seventh angel sounded; and there followed great voices in heaven, saying, The kingdom * of the world is become the kingdom of our Lord, and of His Christ: and He shall reign for ever and ever" (xi. 14, 15).

But this is only the final catastrophe announced in one burst of celestial triumph (see pp. 87, 88). For it was, as we shall presently see, to require seven outpourings of the vials of God's wrath to bring the kingdom of the beast to utter extinction. "And the four and twenty elders, which sit before God on their thrones, fell upon their faces, and worshipped God, saying, We give Thee thanks, O Lord God, the Almighty, which art and which wast, and which art to come; † because

---

* So both the MSS. and the versions read.

† This last clause, omitted in the R.V., is, I think, sufficiently vouched for.

Thou hast taken Thy great power, and didst reign. And the nations were wroth, and Thy wrath came, and the time of the dead to be judged (see vi. 10), and to give their reward to Thy servants the prophets, and to the saints, and to them that fear Thy name, the small and the great; and to destroy them that destroy the earth" (xi. 16—18)—in other words, to give redress of all *public* (antichristian) wrongs.*

"And there was opened the temple of God (the 'sanctuary'—ναός) that is in heaven; and there was seen in His temple the ark of His covenant; and there followed lightnings, and voices, and thunders, and an earthquake, and great hail" (xi. 19).

But why did the temple require opening? for we found (xi. 1) that the sanctuary (the

---

* When we find language like this—of "judging" enemies of the truth, and "rewarding" the faithful—we must not immediately conclude that the "time" referred to is "the day when God shall judge the secrets of men by Jesus Christ," "the judgment of the great day." We should always consider whether the party spoken of is particular *individuals* or *public bodies* or *systems*. As these can only be judged in this life, and on the theatre of the present world, the "judgment" of them can only be the complete overthrow of them. Yet this is far too much overlooked.

ναός) was not only open, but the altar and worshippers were seen in it. The answer is, that it had been daringly shut by *the Church*, which taught that, instead of every believer "having boldness to enter into the holiest by the blood of Jesus" (Heb. x. 19), there is required the intervention of a human priest in the transactions between the soul and God. No wonder that the "searching of the Scriptures," which teaches the reverse of this, is frowned upon and all but legally prohibited! "Woe unto you, scribes and Pharisees, hypocrites! for ye shut up the kingdom of God against men: for ye enter not in yourselves, neither suffer ye them that are entering to go in." But the year of release has come, and the slaves are free. And the concussion of the elements—the lightnings, and voices, and thunders, and the earthquake, and great hail—are the symbolic expression of the revolutionary crash of this mighty movement.

Such, then, is the grand catastrophe; but it is only *announced* in one brief outburst of triumph. For the kingdom of the beast has been too slowly and solidly built up, and of too long duration, to quickly disappear. Shaken to its centre we have seen it to have been by the earthquake of the sixteenth century; and what it has lost it will never recover. But the Seven Vials that fill up the wrath of God against this dragon-inspired Power have yet to be poured out.

THE SEVEN VIALS.

"And I saw another sign in heaven, great and marvellous: seven angels having seven plagues, which are the last; for in them is finished the wrath of God. And I saw as it were a glassy sea mingled with fire; and them that come victorious from the beast, and from his image, and from his mark, and from the number of his name, standing on the glassy sea, having harps of God" (xv. 1).

This word-picture is singularly expressive both of their untroubled safety and the fiery trial through which they reached it. It is the noble army of martyrs who, during the long reign of that blasphemous Power, counted not their lives dear unto them, but cheerfully surrendered them rather than deny the Lord that bought them.

THE CHORAL HYMN.

"And they sing the song of Moses and the song of the Lamb."

Yes, "the song of Moses": "The enemy said, I will pursue, I will overtake, I will divide the spoil: my lust shall be satisfied upon them. I will draw my sword, my hand shall devour them. Thou didst blow with thy wind, the sea covered them: they sank as lead in the mighty waters" (Exod. xv. 9, 10). "So let all Thine enemies perish, O Lord: but let them that love Thee be as the sun when he goeth forth in his might" (Judg. v. 31).

"And the song of the Lamb": for "they overcame by the blood of the Lamb, and by the word of their testimony; and they loved not their lives even unto death" (xii. 11). But for this *final* victory they have a special note of praise, saying, "Great and marvellous are Thy works, O Lord God, the Almighty; righteous and true are Thy ways, Thou King of the nations.* Who shall not fear Thee, O Lord, and glorify Thy name? for Thou only art holy; for all the nations shall come and worship before Thee, for Thy righteous acts have been made manifest" (xv. 3, 4). The allusion here, I think, is to the breaking up of the whole complicated system of antichristianism—*Church* and *State*; each working into the other's hands, each for its own ends, and both against Christ.

"And after these things I saw, and behold the temple of the tabernacle of the testimony in heaven was opened: and there came out from the temple the seven angels that had the seven last plagues, clothed with fine linen,† pure and bright, and girt about the breasts with golden girdles. And one of the four living creatures gave unto the seven angels seven golden vials full of the wrath of God, who liveth for ever and ever. And the temple

---

* There are three readings of this clause: "King of saints" (received text and A.V.); "King of the ages" (R.V.); and "King of the nations." This last is, I think, best supported, and best suits the context.

† Not (as in the R.V.) "arrayed with [precious] stone, pure and bright"—one of the very worst readings, of which have elsewhere written. (See Addendum II., p. 219.)

was filled with smoke from the glory of God, and from His power; and no one was able to enter into the temple, till the seven plagues of the seven angels were finished" (xv. 5—8).

The general import of this seems to be that, till the air could be cleared of the noxious worship that had been offered there by the followers of Antichrist, those who "offer spiritual sacrifices, acceptable to God by Jesus Christ," could not be seen there.

To take up each of these last Septenaries in detail, is no part of my plan in this A B C of the Apocalypse; but only to do as with the *Seals* and the *Trumpets*,—to indicate what seem to be the turning-points of historical prediction which, as we judge, they contain. It is a succession of "plagues" to befall the kingdom of Antichrist, in one feature of it or another. Some of them are very significant. Thus, the third is (like the first of the plagues of Egypt) "turning the waters into *blood*." Yes, says the angel of the waters, and "Thou art righteous, O Lord, who hast thus judged: for they shed the blood of saints and prophets, and blood hast Thou given them to drink, for they are worthy. And I heard the altar" *—a voice, it would seem, from "the souls of them that had been slain, for the Word of God and for the testimony which they held" (not now against *Paganism*, as in vi. 10, but against *Antichrist*)—" Yea, O Lord God, the Almighty, true and righteous are Thy judgments" (xvi. 4—7).

"The fifth angel poured out his vial upon the throne of the beast"—the seat of supreme ecclesiastical authority; "and his kingdom was full of darkness"—like the ninth, and all but the last, of the plagues of Egypt. What this darkness may be must be left to the events themselves to tell. But the effect— "And they gnawed their tongues for pains, and blasphemed the God of heaven, and repented not of their works"—that we can well understand.

One obstacle only, to the capture of " Babylon" and the fall of the "throne of the beast," remained to be taken out of the way; and this is predicted in the sixth vial :—

"And the sixth angel poured out his vial

---

* So read the best authorities.

upon the great river, the river Euphrates; and the waters thereof were dried up, that the way might be made ready for the kings that come from the East (Gr. 'the sun-rising')" (xvi. 12).

The river Euphrates was the glory of the literal Babylon, and made it impregnable, until Cyrus and the princes that followed him turned its course and dried up its waters. Having thus entered it dry-shod, Babylon fell, and the kingdom with it. And thus the last obstacle to the fall of mystical Babylon and the kingdom of Antichrist — whatever that may be—will be taken out of the way.*

"And I saw coming out of the mouth of the dragon, and out of the mouth of the beast, and out of the mouth of the false prophet, three unclean spirits, as it were frogs: for they are spirits of demons, working signs; which go forth unto the kings of the whole world, to gather them together to the war of the great day of God, the Almighty" (xvi. 13, 14).

---

* Those who take the "drying up of the river Euphrates" to point to the fall of the Turkish Empire, and "the kings of the East" to be the Jews returning to their own land go very wide of the mark.

All the malignant elements of the carnal mind, which is enmity against God, seem here represented. Whatever comes from the mouth of the *dragon* must be the diabolical spirit, pure and simple ; that from the mouth of the *beast* (as distinguished from "the false prophet") must be the *civil* arm of Antichrist, that which carries into execution the sentences of the head of the Church, shedding the blood of the saints ; and that which comes out of the mouth of the false prophet can be no other than the claims of "the man of sin and son of perdition," who "opens his mouth in blasphemies against God, to blaspheme His name, and His tabernacle, and them that dwell in heaven." All these malignant spirits, which were "like frogs" (unclean, noisy, offensive creatures) "are the spirits of demons," * taking active pos-

---

* It is a pity that the New Testament revisers have retained from the A.V. the word "devils" in this and all other places where the word occurs. The Greek word for "devil" is never used in the New Testament, but always a diminutive of the word "demon." A good deal is lost by this. Thus, in James ii. 19, "Thou believest that there is one God: the *demons* believe, and tremble"—an express reference to the shriek which they gave at the sight of their future Tormentor:—"What have we to do with Thee, Jesus, thou Son of God? art Thou come to torment us before the time?" (Matt. viii. 29). The distinction

session of those who are their willing tools, and inspiring them with hostility to Christ.

"Working signs:" compare 2 Thess. ii. 9, 10, —Even he, whose coming is according to the working of Satan, with all power and signs and lying wonders, and with all deceit of unrighteousness for them that are perishing; because they believed not the love of the truth, that they might be saved."

"Which go forth unto the kings of the whole world, to gather them together to the war of the great day of God, the Almighty" (xvi. 14). A gigantic confederacy against Christ.

"Blessed is he that watcheth, and keepeth his garments, lest he walk naked, and they see his shame" (xvi. 15). This is a note of preparation for the final catastrophe; and, as I judge, a call to such of "God's people" as are still in "Babylon" to come out of her in time, lest they be involved in her ruin (see on xviii. 4).

"And they gathered them together into the

---

seems to be this: when the word "devil" is used, it is always in connection with *sin*, as tempting to the commission of it, and stimulating the sinful principle; whereas, when one is *possessed* by a "demon," he is for the time not his own conscious self, but is the organ of one who *through him* is the speaker and actor.

place which is called in Hebrew Har-Magedon" (xvi. 16). The allusion seems to be to Megiddo, where Josiah was slain (2 Chron. xxxv. 22; and compare Zech. xii. 11). The details of this last war are *reserved*, to be taken up in chap. xix. 11—21.

Now comes the announcement of the final catastrophe: "And the seventh angel poured out his vial upon the air; and there came forth a great voice out of the temple, from the throne, saying, It is done. And there were lightnings and voices and thunders: and there was a great earthquake, such as there was not since men were upon the earth, so great an earthquake and so mighty. And the great city was divided into three parts, and the cities of the nations fell,"—all expressive of such a revolutionary crash as would leave nothing of what had been the most outstanding and seemingly enduring features of the former condition of Church and State.

"And Babylon the great was remembered in the sight of God, to give unto her the cup of the wine of the fierceness of His wrath. And every island fled away, and the mountains were not found. And great hail, every stone about the weight of a talent, cometh down out of heaven upon men : and men blasphemed God because of the plague of the hail; for the plague thereof was exceeding great' (xvi. 19—21).

Now, for the first time, the great city which spiritually is called *Sodom* and *Egypt* and *Jerusalem* (the slaughterhouse of the prophets, " where also the Lord was crucified "—xi. 8), gets the name of "BABYLON the Great,"—its time of doom has at length come into remembrance in the sight of God; and the cry is heard, " Fallen, fallen is Babylon!" (xviii. 2).

Thus closes the last of the Septenary series —that of the Seven Vials; and with it all that Divine Wisdom has seen it fit to announce beforehand of the fate of that gigantic antichristian edifice of Church and State, the features of which, so fully described in preceding chapters, we have pointed out in detail.

But, as if to make assurance doubly sure as to what this accursed system is, a *key* to the whole mystery is given in—

Chap. xvii.
THE KEY TO THE MYSTERY.

"And there came one of the seven angels which had the Seven Vials, and spake with me, saying, Come hither, I will shew thee the judgment of the great harlot that sitteth upon many waters"—that is, as explained in verse 15, "peoples and multitudes and nations and tongues" (the consolidated fragments of the broken empire)—"with whom the kings of the earth have committed fornication, and they that dwell in the earth (their subjects) have been made drunk with the wine of her fornication. And he carried me away in the Spirit into a wilderness: and I saw a woman sitting upon a scarlet-coloured beast, full of names of blasphemy, having seven heads and ten horns" (xvii. 1—7).

Yes: now we have it out. The "woman" of chap. xii. 1, once the beauteous young "bride, the Lamb's wife," alas, how changed! Raised to honour in the empire, she that was so long persecuted is now patronised and petted—caressed until she yields to the embraces of a stranger, and has become a harlot! And what stranger she embraced the symbol itself tells us: it is the beast with seven heads and ten horns, but the diadems no longer on the *heads* (xii. 3), but on the *horns* (xiii. 1). For, says the angel, it is "*the kings of the earth*" that commit fornication with her. But since these "kings" had no existence till "the heads" ceased to exist on the fall of the Empire, this fixes the sense to be that of the "kings" that followed the breaking up of the Empire—the "horns" that had the diadems when there were no "heads."

"And the woman was arrayed in purple and scarlet, and decked in precious stones and pearls"—the sumptuous gifts of the kings to her, in return for what she did for them. Thus beautifully did *Church* and *State* play into each other's hands—the *kings*, including "them that dwell on the earth" (the rich laity), vying with each other who would do most honour to the *Church*; providing for it on a scale corresponding with the grades of rank and wealth in the *State*; while she, on the other side, compacts them—though each independent of the other—into one solid Christendom,—a gigantic engine of Church and State, which for more than a thousand years, as *the Church* condemned to death as "heretic" all who dared to refuse its claims, while as *the State* its sentences were dutifully executed.

What do I see in this? I see reproduced to the life the story of Herodias and Herod Antipas in their treatment of John the Baptist. When that holy man dared to say to Herod, "It is not lawful for thee to have her," from that moment his head was forfeited. Herod would have stood this; for he had a conscience, and "did many things, and heard John gladly." But that wretch with whom he lived had no sense of shame, and was determined to have his life. Herod compromised the matter by imprisoning the holy man. But so long as he lived to trouble her royal paramour she could not rest; and at length she found her opportunity. At a great birthday feast, with all the high magnates of the kingdom around him, Herodias had her daughter brought in to dance lasciviously before them. And so fascinated was the king that, in a hasty moment— perhaps in his cups—he promised her with an oath whatsoever she might ask, to the half of his kingdom. "What am I to ask?" said the poor girl to her mother. "The head of John the Baptist," was the instant reply. She herself could not touch a hair of his head, but she had the king now in her fix. He was stung by the unexpected demand; but he had passed his word, and it had to be done. The keep of the castle where the feast was held was no doubt the place of John's imprisonment—within a few yards of the banquet hall. The executioner therefore would quickly do his business; and, returning with the bleeding head in a charger, give it to the girl, and she to her mother. And we can imagine how that vile woman would gloat over the ghastly spectacle.

Even so, our "harlot," though she could not herself "shed the blood of the saints and of the martyrs of Jesus," could cause the execution of her cruel sentences by "the kings"—by an *auto da fe* (an "Act of Faith"). No wonder, then, that she was seen "drunken with their blood."

"Having in her hand a golden cup, full of abominations, and the unclean things of her fornication. And upon her forehead a name written, MYSTERY, MYSTERY, BABYLON THE GREAT, THE MOTHER OF THE HARLOTS, AND OF THE ABOMINATIONS OF THE EARTH. And I saw the woman drunken with the blood of the saints and of the martyrs of Jesus. And when I saw her I wondered with great wonder. And the angel said unto me, Wherefore didst thou wonder? I will tell thee the mystery of the woman, and of the beast which carrieth her, which hath the seven heads and the ten horns. The beast that thou sawest was, and is not, and is about to come up out of the abyss and to go into perdition" (xvii. 4—8).

"The beast" is the seven-headed and ten-horned dragon of chap. xii. 3; and "it *was*" while the diadems were on the "*heads.*" But it "was not," when the heads (or what they meant) ceased to exist, on the fall of the Empire. But it "was about to come up out of the abyss," when it appeared in a new form—wounded to death, but the wound healed—a blasphemous and bloody form, inspired by the great red dragon to drink the blood of the saints, and of the martyrs of Jesus; but "to go into perdition."

"And they that dwell on the earth shall wonder, they whose names have not been written in the book of life from the foundation of the world, when they behold the beast, how that he was, and is not, and is to come."

Though the language here employed would seem to take in its sweep all but those who are to be saved, it is evident from chap. xiii. that only *those who have identified themselves with the system*, its originators, manufacturers, upholders and agents in its blasphemies and cruelties, are included. Let any one read the description of them in 2 Thess. ii. 10—12, and he will see how inapplicable that is to vast numbers of the nominal adherents of antichristian Romanism. So far from such *not* having their names in the book of life, we shall find when we come to chap. xviii. that God will have a people whom He calls His own in "Babylon" almost to its fall. But the class here meant is a totally different class. And of them we must learn to think as God Himself and all holy men do. Can we but execrate men who, as members of "the Holy Inquisition," for example—established by the supreme authority of the Church to try "heretics"—could year after year doom to death, in every form of cruelty, some of the noblest characters, including ladies of the highest distinction, in cold blood and in the name of religion? In view of such things, one is constrained to say, with the Psalmist,

"Do not I hate all them that hate Thee? Can I but be grieved with them that rise up against Thee? I hate them with perfect hatred: I count them mine enemies" (Ps. cxxxix. 21, 22).

"Here is the mind which hath wisdom. The seven heads are seven mountains on which the woman (the harlot) sitteth,"—the city of the seven hills (*urbs septicollis*): "and they are seven kings (or forms of government); the five are fallen: the one is, and the other is not yet come: and when he cometh, he must continue a little while. And the beast that was, and is not, is himself also an eighth, and is of the seven; and he goeth into perdition" (xvii. 9—11).

The mystery of this puzzle is not so great as at first it seems to be. If five of the different forms of government had already fallen, and the one that then existed was that of Emperors, the other, not yet come, must be that which *succeeded the fall* of the Empire. This is "the beast that *was*," in its *Imperial* form as a persecuting power, but "*is not*" when the Empire fell; "he is himself an eighth, and is of the seven" (an *eighth*, inasmuch as, being an ecclesiastical power, it was perfectly different from all the others; but "of the *seven*, since the same *supremacy* which characterised all the seven forms was prolonged in this one): "and he goeth into perdition," being that blasphemous, persecuting Power, "drunken with the blood of the saints."

"And the ten horns which thou sawest are ten kings, which have received no kingdom as yet (while the Empire stood), but they receive authority with the beast for one hour (or contemporaneously with the beast). These have one mind, and they give their power and authority unto the beast (to execute the sentences of death which he pronounces upon the 'heretics'). These shall war against the Lamb, and the Lamb shall overcome them; for He is Lord of lords and King of kings; and they that are with Him are called, and chosen, and faithful."

Beyond doubt this is a religious war—a last one, and on a vast scale. But this (xvii.) being only an *explanatory* chapter, the *details* of it are reserved, and will be found in chap. xix. 11 to the end.

"And he saith unto me, The waters which thou sawest, where the woman sitteth, are peoples and multitudes and nations and tongues,"—the nations of Christendom (xvii. 12—15).

But what becomes of "the harlot"? She shares the fate of all harlots—is hated by the powers that seduced her with a hate proportioned by the love—the lust wherewith they lusted after her (as in the case of Amnon with his sister Tamar—2 Sam. xii. 15).

"And the ten horns which thou sawest, and* the beast, these shall hate the harlot." Let the reader carefully note how "the beast," being a compound power—ecclesiastical and secular combined, as a great engine (*condemning* in its ecclesiastical capacity and *executing* secularly the sentences against the "heretics"), is sometimes viewed separately in its ecclesiastical capacity, and sometimes in its secular. Here it is viewed in its *secular* capacity, and as such it hates the harlot. "These shall kill the harlot, and make her desolate and naked (stripping her, I suppose, of the trappings they gave her), and shall eat her flesh,

---

* Not "upon the beast"—as the received text and A.V.

and burn her utterly with fire." Under the Jewish law, "the daughter of any *priest*, if she profane herself by playing the harlot, she profaneth her father, she shall be burned with fire" (Lev. xxi. 9). The significance of this allusion is obvious.

"For God did put in their mind to do His mind, and to come to one mind, and to give their kingdom (their sovereign authority) unto the beast, until the words of God should be accomplished."

"And (in a word) the woman whom thou sawest is the great city which reigneth over the kings of the earth" (xvii. 16—18).

Thus ends this explanatory chapter—the "Key," as I have well called it—to the whole symbolical prediction of the great antichristian Power, that has held us so long.

We are now in possession of all the materials for fixing—if that can be done—the time when the famous 1260 years *closed*, and of course also when it *began*. Everything depends (I need not say) on the accuracy of our interpretation of the symbols, and of the action of the parties symbolised. But supposing our exposition, in its leading features, to be correct, I now proceed to apply them (with unfeigned diffidence) to the question now before us.

I have said, then, that all that is predicted of the authority given to "the beast" to "continue for forty and two months," the "worship" paid him throughout this period, and his "war" with the saints, his "overcoming" and "killing" them till they were but Christ's "two witnesses," and even they at length killed—none daring to take up (as John's disciples did to his corpse) their dead bodies to bury them—all this, I have said, expresses to me the period, not of his *existence*, but of his *unimpaired supremacy*. Accordingly, I am constrained to date the close of the symbolical "forty-two months" at the time when this supremacy was thoroughly broken —when "the earthquake" caused the tenth part of the city to fall, and there were "slain" symbolically "seven thousand" of his adherents, and he could no longer boast of the *unity* of the Church as one of the marks of a true Church which she alone possessed. That time, I said, was undeniably when the great rupture of the sixteenth century took place.

But (you will say) does not this throw the *beginning* of the 1260 years too far back into the earliest Church history? So I long thought, and regarded it as fatal to the supposed *close* of this period. But, when I went back to the exegesis of the text, and found it yielded no other result than that given in the preceding pages, I asked myself why we should expect the *beginning* of this period to be marked so much by *unmistakable outstanding events*, and whether the first rise of the great antichristian power might not be found to lie at the *hidden springs of the Church's defection from Christ*.

Let it never be forgotten that the two great antagonistic and irreconcilable forces that press upon the heart of man are *the Church* and *the world*—in the biblical sense of these terms. Neither of these can live under the dominant influence of the other. "Ye (said Christ to His genuine disciples) are the light of the world. . . . Ye are the salt of the earth,"— to illuminate and saturate society with your principles and character. Living Christianity is essentially aggressive; but so its opposite. "If the light (then) that is in us is darkness, how great is that darkness! and if the salt has lost its savour, wherewith shall it be salted? It is thenceforth good for nothing, but to be cast out, and trodden under foot of men."

So long as the ministers of Christ, as keepers of "the keys," are jealous of the Church's purity, and have the courage of their convictions to keep it pure, their Master will own and honour them. But if they *let the world into the Church*, its outward form may remain unchanged, but its distinctive character is lost—all that is *vital* in it dying down, and the way thus prepared for corruptions of every sort. In the Vision of chap. xii.—where the Church is seen in all the beauteous vigour of her youth, as "the Bride the Lamb's wife"— "the dragon," that old serpent, well knowing the key that most easily opens the clerical heart, begins with the ministers of the Church, the "stars of heaven" (i. 20), "a third part of whom with his tail (stealthily) he casts down to the earth." His key to them was *ambition*, which, like the horseleach, is never satisfied. Bishops became archbishops; then came metropolitans, patriarchs, primates—with corresponding influence and wealth—culminating inevitably in the See of Rome: a See which in persecuting times marked out its occupant as the first to be the victim of the persecuting edict; but when Christianity, in the person of Constantine, sat on the throne of the Empire, the most coveted of all posts in the Church.

All this, *full-blown*, we have not in the Apocalypse (as we have seen), but in two astounding predictions of the apostle Paul— 2 Thess. ii. and 1 Tim. iv.—which not only confirm all that we have found in the Apocalypse, but show with what rapidity, when the tide once set in, these things would be realised. I must take them up therefore in detail.

"Now, we beseech you, brethren, touching the coming of our Lord Jesus Christ, and our gathering together unto Him, that ye be not quickly shaken from your mind, nor yet be troubled, either by spirit (pretended revelations), or by epistles as from us (forged letters), as that the day of the Lord is [now] present (already come). Let no man beguile you in any wise: for [it will not be], except the apostacy (or 'the falling away') come first, and the man of sin be revealed, the son of perdition, he that opposeth and exalteth himself against all that is called God or that is worshipped. So that he sitteth in the temple (or 'sanctuary'—εἰς τὸν ναὸν) of God, so that he* setteth himself forth as God. Remember ye not, that, when I was yet with you, I told you these things? And now ye know that which restraineth, to the end that he may be revealed in his own season. For the mystery of lawlessness doth already work: only [there is] one that restraineth now, until he be taken out of the way. And then shall be revealed the lawless one, whom the Lord Jesus shall slay with the breath of His mouth, and bring to nought by the manifestation of His coming; even he, whose coming is according to the working of Satan with all power and signs and lying wonders, and with all deceit of unrighteousness for them that are perishing; because they received not the love of the truth, that they might be saved. And for this cause God sendeth them a working of error, that they should believe a lie, that they all

---

* The words in the A.V. (from the received text)—"as God"—are not in the true text.

might be judged (condemned), who believed not the truth, but had pleasure in unrighteousness" (2 Thess. ii. 1—12).

No less significant, though less extended, is the other prediction, in 1 Timothy :—

"Now the Spirit (by the mouth of some prophet) saith expressly, that in later times some shall fall away from the faith, giving heed to seducing spirits and doctrines of demons, through the hypocrisy of men that speak lies, branded in their own conscience as with a hot iron; forbidding to marry, and commanding to abstain from meats, which God created to be received with thanksgiving by them that believe and know the truth. For every creature of God is good, and nothing to be rejected, if it be received with thanksgiving: for it is sanctified through the word of God and prayer" (1 Tim. iv. 1—5).

Let us now try to gather up the substance of these pregnant predictions.

Since in his First Epistle the Second Coming of our Lord had been made so prominent, we need not wonder that such raw converts as the Thessalonians should get excited about it, should misunderstand the teaching they had received, and be open to every kind of wild and unsettling expectations on the subject. In his Second Epistle, while adverting to this state of things, it might have been enough to tell them how mistaken they had been in supposing that the time for Christ's Second Coming had already arrived, and that great events had to take place before that event. But instead of that he is led—no doubt by the inspiring Spirit—to make so explicit a revelation of what events the Church was to expect, that we must note and comment on them in connection with his other equally explicit prediction to Timothy.

The first great event to be looked for was "the apostasy," but not that gradual decay in the spiritual life and love of the early Churches, which might have been expected, and is to be seen at times in the Churches of our own day. Besides that it is emphatically called "*the* apostasy," the descriptions given of it make it plain that it was to be no ordinary "falling away." It was to be such an outbreak of clerical ambition as there was no conceivable field for when the Christians in their little gatherings called churches were a persecuted people, and

Christianity in the Empire a prohibited religion. It supposes a state of things the reverse of all this—Christianity and the Christian churches in possession of full liberty, *the Church* an organised body, and its ministers able to mount to heights of assumption almost incredible, and all this unopposed. Has this been realised? Let the known facts give the answer. Did not the Bishop of Rome at length claim to have supreme authority in the Church of Christ, and be the fountain of all ecclesiastical power—*the Vicar of Christ* upon earth? And, instead of this being resisted by his subordinates, and in course of time becoming a dead letter, did not the largest conclave of bishops that could be assembled at Rome but a few years ago vote the Pope, when he speaks *ex cathedrâ*, to be, in matters of doctrine, *personally infallible*—his utterances in such cases being *the voice of God*? Yes, this man is a king of kings; for he wears a triple crown—the diadems of other sovereigns not being distinctive enough for him. If this does not come up to all that is here predicted of "the man of sin," we may safely say that the prediction will never be realised.

"Lawlessness," again, is a distinctive feature of

this "man of sin." He is "*the* lawless one" —not so much, I think, that he disregards laws, like rebels and traitors, but that he sets himself above all law, except his own, which he imposes upon his subjects (the office-bearers and members of the Church) in daring opposition to the law of God. If it be asked how he does this, I refer for answer to 1 Tim. iv.

"*Forbidding to marry.*" The reference here is not to the mere creeping in of an *ascetic spirit*, that spurious spirituality which consists in self-imposed mortification of the flesh, and which any one acquainted with the religious workings of the human mind might predict as likely to arise in the Church. It is this spirit *organised into a system and worked out by Church authority*, in the specific form of a prohibition of marriage.

Now, in point of fact, there is one, and only one, such body existing. *The Church of Rome forbids clerical marriage*, and holds celibacy forth as a holier state of life, and therefore the fitting thing for those who minister in holy things. But He who made man said, "It is not good for man to be alone," and therefore He made woman. In some cases, of course, a single life is best, and may become a duty. But Divine authority we have for saying that "marriage is honourable in all" (Heb. xiii. 4).

Enforced celibacy cuts off a large and very influential portion of society from social life. To all the sympathies, the tenderness, the exquisite delicacies which make the sweetness of family life, they are, in personal experience, entire strangers. And what must be the effect of cutting them off from all this *by a law*, but to drive them insensibly into a penfold of their own order, and (by almost necessary consequence) to make their own interest, as an order, paramount with them, and especially in all that tends to enrich *the Church*? When, in our Lord's time, some pious Jew dedicated to the Temple part of his means, but afterwards, when his means were gone, and he became unable to support his aged and destitute parents, he would wish to withdraw part of his gift to the altar in obedience to the higher law, the "scribes and Pharisees, hypocrites," had but to utter the word "Corban" (gifted), and though his parents should starve, no matter, it could not be touched. "The Church," "The Church," their modern successors would say in such cases. To such an extent, in fact, had this grasping of money—all for the Church, no doubt—proceeded, that our own as well as other countries has been forced to interpose by law, to prevent glaring injustice being done in the name of religion.

Nor is this all. For as keepers of their people's consciences (but who made them such?)—every one having his or her Father Confessor—they have it in their power to worm themselves into the most sacred and delicate intercourses of family life and personal character. No doubt it is possible to hear confession with the highest sense of honour and propriety, and many there are who, fully alive to the demoralising tendency of too searching inquiries, discharge what they regard as a clerical duty with unexceptionable propriety. But within the home circle of every family there is much with which those outside of it—whoever they are—have nothing, and ought to have nothing, to do.*

---

* I once travelled in Ireland with an intelligent Roman Catholic gentleman of property. Getting into conversation, and

But if in this case "the lawless one" has presumed to traverse the Divine order of human society, there is another way in which more nakedly he does what the apostle denounces.

Lastly, "commanding to abstain from *meats*, which God created to be received with thanksgiving." I refer, of course, to the "command" to eat no *flesh* meat on one day of the week. Some, indeed, in all our Churches think that on the day on which our Lord was crucified, it would help to deepen their devotion to fast during part of the day—say the three hours' darkness—or to abstain from flesh meat on that day (Friday). To me such self-imposed mortification of the flesh is not like the masculine character of the Apostle's teaching, and what I venture to call the common-sense style in which he here speaks of meats. But,

---

happening to agree in politics, more than either of us supposed, I ventured to ask him whether—if we Protestant clergy, who could marry or lead a single life as we pleased, were forced to live a single life—he did not think that (human nature being what it is) the effect would be to make the interests of our own order paramount, in politics and everything else. "You have touched our weakest point," he said. "It is the curse of our Church; but we daren't speak out."

be this as it may, the thing which the "lawless one" does is to "*command* abstinence from" flesh meat on Fridays—and this in the name of *religion*, which religion consists in what the Apostle here condemns.*

Returning now to 2 Thess. ii.

There is a mysterious, a studied caution in the way in which the Apostle refers to "that which restraineth the lawless one" from being "revealed." "And now ye know that which restraineth. For the mystery of lawlessness doth already work: only [there is] one that restraineth, until he be taken out of the way. And then shall be revealed the lawless one." Now, why all this hesitation in naming the "one that restraineth" the revelation of that dread Power? The early fathers of the Church all dreaded the approach of the *Antichrist*, as they rightly judged that this lawless one would be; and if *Irenæus* speaks for them, the one obstacle to his appearance was the existence of an *Emperor*. But to speak that out nakedly might be dangerous; and therefore (he thinks)

---

* Of the defence of this law—that it is only for one day in the week, and for the best of purposes—it is enough to say that to *command* it at all is what is here condemned.

it was that in the Apocalypse his name is concealed in the numerical puzzle of "six hundred sixty and six" (xiii. 18).

Whatever there may be in this conjecture, certain it is that, until *Imperial* supremacy ceased by the fall of the Empire, *Ecclesiastical* supremacy—binding the separate sovereignties, which were independent of each other, into the one supremacy which they all recognised—was impossible. The Empire fell in A.D. 476. And as we have seen that almost from the accession of Constantine (A.D. 312) as a Christian emperor, the spiritual life of the Church rapidly declined, it is not too much to say that the rise of Antichrist had its *springs* then.

But even granting this (it may be said), would it not throw the close of the 1260 years to a later period in the sixteenth century than *the rupture of the Church's unity* by the movement which Luther began? To which it is enough to reply, that the great breach was a succession of ruptures. In some of the countries—as Holland and Scotland—the breach did not take place till nearly half a century after Luther's movement.

But I confess I am not careful to meet this

kind of objections. What force there is in them I have myself once and again felt, and they have driven me to a fresh examination of the text, not once only. But finding that it would not *to me* yield any conclusions but those I here submit to the reader, I must leave the decision to others. Precise dates I have no skill in, and it well befits a mere A B C to leave these alone.

One difficulty I must notice. If the 1260 years terminated so early as the Reformation of the sixteenth century, does not that make the pouring out of the *Vials* begin far too early, and suppose them to require exhaustion before this time? But is it said that the one was immediately to follow the other? Certainly not. All that the text necessarily implies is, that the *Vial series* would be *the next cycle of events*, the next stage in the march of the great events here predicted. The language employed does, indeed, seem to imply that, when the time, the set time, arrives for the pouring out of the *first* of the vials of God's wrath upon this accursed system, the rest may be expected to follow in quick succession, until its final extinction. But even that is not certain. Perhaps it may be otherwise. For who can tell but that after the first great shock of the first "vial," the Papal power may settle down and recover some of its lost ground? It did this, we know, when the Protestant triumphs at the Reformation period, begetting self-confidence, led their Churches to be more intent on settling controversies amongst themselves than on gaining fresh ground. Such alternate rises and falls on the part of the Papacy are quite consistent with a steady ebbing out of the life and strength of the system. But when the time appointed comes —as come it will—then shall be brought to pass what we find in the Vision of chap. xviii.

THE FALL OF BABYLON.

"After these things I saw another angel coming down out of heaven, having great authority; and the earth was lightened with his glory. And he cried with a mighty voice, saying, Fallen, fallen is Babylon the great, and is become a habitation of demons, and a hold of every unclean spirit, and a hold of every unclean and hateful bird. For by the wine of the wrath of her fornication all the nations are fallen* (probably as drunkards, who cannot stand); and the kings of the earth committed fornication with her, and the merchants of the earth waxed rich by the power of her wantonness" (xviii. 1—3).

It is a fine feature of this book that its darkest scenes are lighted up with poetry of a kind of its own, yet easily understood. What, for example, can be grander than the opening verses of this chapter? And the loathsomeness of the fallen object, how could that be more graphically described? The figure is taken from the description of the fall of *Idumæa* (Isa. xxxiv. 14, 15), and especially Jeremiah's description of the fall of *Babylon* (chap. li.), the whole of which the reader should go carefully over afresh, in the light of the application of its language here made to *mystical Babylon*.

SUMMONS TO GOD'S PEOPLE TO COME OUT OF BABYLON BEFORE ITS FALL.

"And I heard another Voice from heaven, saying, Come forth, My people, out of her,

---

* Such is the reading of the best authorities.

that ye have no fellowship with her sins, and that ye receive not of her plagues" (xviii. 4).

Two things are obviously here : (1) that God will have a people in Babylon, whom He calls "My people," almost to the time of its fall ; (2) that only by "coming out" of her will they be held guiltless of her sins and be excepted from her doom. (Even here the figure is taken from a similar warning to flee out of literal Babylon—Isa. lii. 11 ; Jer. li. 9.) And just as the Lord Jesus gave His disciples a sure sign of the time for them to flee out of Jerusalem before its fall (Matt. xxiv. 15—18, Luke xxi. 20—22), and they did as they were told, and all escaped ; so there can be no doubt that the actual escape from Babylon before its fall of all that God regards as "My people" will be seen to in that day.

And when we know how every Roman Catholic is taught to believe that "out of *the Church* there is no salvation," and that this means out of the Church of Rome ; and how even devout and intelligent Romanists can hardly persuade themselves but that separation from Rome means separation from Christ—who can wonder that many will be found of God's "people," till all but the very last? Readers of Church history know that such distinguished theologians, scholars, and blameless Christians as *Cardinal Contarini* and *Cardinal Sadolet* were of one mind with Luther and Calvin on the doctrine of justification and kindred truths, but could not see their way to break with Rome. And a clerical brother who travelled as tutor to the son of a wealthy banker over almost every city of Europe, and conversed with many intelligent and excellent priests, told me that many of them were genuine Protestants in all but the name.

"For her sins have reached even unto heaven, and God hath remembered her iniquities. Render unto her even as she rendered, and double unto her the double according to her works: in the cup which she mingled, mingle unto her double. How much soever she glorified herself, and waxed wanton, so much give her of torment and mourning: for she saith in her heart, I sit as a queen, and am no widow, and shall in no wise see mourning." This is an adaptation of Isa. xlvii. 8—11 regarding literal Babylon to its facsimile in the mystical. "Therefore in one day shall her plagues come, death, and mourning, and famine; and she shall be utterly burned with fire, for strong is the Lord God which judged her" (xviii. 4—8).

A DIRGE OVER THE FALL OF BABYLON.

This is an exquisite device—and there are many such—of a book incomparable no less for its songs of "lamentation and mourning and woe" than for its music of heaven. But it is no part of my plan to take up the details of this *Lament*. Three things, however, I must notice in it. First, the stately *swing* of its music (remarkably well brought out in our version). Next, that *pause* in the midst of the wail, to tell us (verse 21) how, like as the prophet Jeremiah caused his servant to read to King Zedekiah his prediction of the fall of Babylon, and then, binding a stone to it, to cast it into the river Euphrates, saying, "Thus shall Babylon sink, and shall rise no more again" (Jer. li. 59—64), so here "a strong angel took up a stone, as it were a great millstone, and cast it into the sea, saying, Thus with a mighty fall shall Babylon, the great city, be cast down, and shall be found no more at all." Lastly, the wail is followed with the bill of indictment against her, in which even the blasphemies of this antichristian "beast" have no place, but her protracted and accumulated cruelties, as the great persecutor of all the prophets and martyrs upon earth are held forth to view: "*And in her was found the blood of prophets, and of saints, and of all that have been slain upon the earth*" (verse 24).

HALLELUJAHS IN HEAVEN OVER THE FALL OF BABYLON.

Chap. xix. 1—9.

"After these things I heard as it were a great voice of a great earthquake, Hallelujah! Salvation, and glory, and power, belong to our God: for true and righteous are His judgments; for He hath judged the great harlot, which did corrupt the earth with her fornication, and He hath avenged the blood of His servants at her hand (according to His promise to the martyrs of the Pagan persecution, vi. 11). And a second time they say, Hallelujah. And her smoke goeth up for ever and ever. And the four and twenty elders and the four living creatures fell down and worshipped God that sitteth on the throne, saying, Amen; Hallelujah. And a voice came forth from the throne, saying, Give praise to our God, all ye His servants, ye that fear Him, the small and the great. And I heard as it were the voice of a great multitude, and as the voice of many waters, and as the voice of mighty thunders, saying, Hallelujah: for the Lord God, the Almighty reigneth. Let us rejoice and be exceeding glad, and let us give glory unto Him: for the marriage of the Lamb is come, and His wife hath made herself ready. And it was given unto her that she should array herself in fine linen, bright and pure: for the fine linen is the righteous acts (δικαιώματά, not δικαιοσυνη) of the saints."

But what is this "marriage of the Lamb" and "His wife"? It is of some consequence rightly to apprehend this. First of all, *betrothal*, by the Jewish law, was legal *marriage*; consequently our Lord was the legal "son of Joseph," though not his actual son. A touching allusion to this sense of marriage was on one occasion made by the Baptist. Some tried to make him jealous of his Master's popularity: "You send disciples to *him*, and at this rate you will soon have none of your own." "You bring me good news," is his noble reply. "Ye yourselves bear me witness that I said, I am not the Christ, but that I am sent before Him. *He that hath the bride is the bridegroom.* But *the friend of the bridegroom,* who standeth and heareth him, rejoiceth greatly, because of the bridegroom's voice. This my joy therefore is fulfilled: He must increase, but I must decrease." Would ye have me step into my Master's place? The bride is not mine: why, then, should the people stay with me? Mine it is to point them to Him. Enough for me to be "the friend of the Bridegroom," sent to *negotiate the match*, bringing together the parties to be wedded; and if I may be privileged to "stand and hear the Bridegroom's voice," and witness the blessed espousals, I seek no more.

But the "marriage" *here* is a step in advance of this. The espousals to which the Baptist refers is that of which the Apostle says to the Corinthians, that he did the same to them as the Baptist did in his day: "I espoused you to one Husband" (2 Cor. xi. 2). But "the marriage of the Lamb" *here* is considerably *in advance of this*; for the "fine linen" with which the bride is "clothed" is "the righteous acts of the saints"—or the steady fidelity to their true Husband of the whole company of the saints in the cloudy and dark day of bitter persecution.

Yes, this is she whom we first met in xii. 1, in her fresh bridal beauty, newly "espoused to one Husband," and now found true to Him, through all the harlotry of the pretended "spouse," and now owned and welcomed by Him, ready to give birth to a fresh progeny of saints, whom "He shall make princes in all the earth" (Ps. xlv. 16).

"And he saith unto me, Write, Blessed are they who are bidden to the marriage supper of the Lamb,"—"Happy they who shall witness such a sight." "And he saith unto me, These are the true words of God. And I fell down before his feet to worship him." So transported was the seer at this spectacle that he must have been almost beside himself, to have been ready to worship the angel who held it up to him. But the rash act is speedily checked by a gentle rebuke: the angel humbly letting him know that he was on no higher level than himself—a fellow-servant with him and all who hold the testimony of Jesus; and God alone he must worship. What a damning rebuke does this administer to that Church which not only directs the same worship to creatures which the angel forbade to be offered to any but God, but sets up in defence of it a fantastic distinction between a lawful worship called *doulia*, and *latria*, the latter of which they reserve to God only! "For (adds the angel) the testimony of Jesus is the spirit of prophecy"—"You are so transported at the spectacle I have held up to you, that you forgot yourself. But in this I have been doing only a prophet's work, holding up Him who is the burden of all prophecy."

THE LAST WAR, AND THE ROUT AND RUIN OF THE PUBLIC ENEMIES OF CHRIST'S KINGDOM.

Chap. xix. 11—21.

We have here at length the battle or "war of Armageddon" (xvi. 16).

"And I saw the heaven opened; and beheld a white horse, and He that sat thereon, called Faithful and True; and in righteousness He doth judge and make war. And His eyes were a flame of fire (the fire of His jealousy for the interests of His kingdom), and upon His head sat many diadems (royal crowns); and He hath a name written, which no one knoweth but He Himself." (Compare His own words, in Matt. xi. 27—" No one knoweth the Son but the Father; neither doth any know the Father save the Son, and he to whomsoever the Son willeth to reveal Him.") Here the sense seems to be that in *this* name—known only to Himself—are wrapt up all that He came to be and do, as His Father's Servant, both for the establishment of His kingdom among men, and the condign punishment of all who dare to oppose it.

"And He is arrayed in a garment sprinkled with blood"—certainly not His own blood, but the blood of His enemies. The whole *idea* and the language is that of the prophet Isaiah, where, in one of his grandest passages, he depicts the fall of *Idumæa*: "Who is this that cometh from Edom, with dyed garments from Bozrah? . . . I that speak in righteousness, mighty to save. Wherefore art thou red in thine apparel, and thy garments like him that treadeth in the winefat? I have trodden the winepress alone. . . . I trod them in mine anger, and trampled them in my fury; and their blood is sprinkled upon my garments, and I have stained all my raiment " (Isa. lxiii. 1—3).

"And His name is called The Word of God.*

"And the armies which were in heaven followed Him upon white horses, clothed in fine linen, white and pure." So far as in the *symbol* the armies issued out of heaven, following their celestial Leader, they were "the principalities and powers in the heavenly places." But since undoubtedly the war is one fought here on earth, and in human flesh and blood, the army that "followed" are the same as in the explanatory chapter (xvii. 14) are

---

* The studious reader should note two interesting facts about the name here given to our Lord. (1) By none of the New Testament writers save the author of the Fourth Gospel and that of this book is this title *as a proper name* given to Him (one of the many proofs that the author of both books is the same). (2) Even in the Fourth Gospel it never occurs in the narrative part, but only in the prologue, to let the reader know, at the outset, that He of whom it treats, though *man*, of our own flesh and blood, was no mere man, but co-eternal and co-essential with God, and Himself God, who in the fulness of time "was made flesh," which was given Him here to do. And when we find this name reappearing here as a proper name of Him who comes forth to execute "the judgment written" against the enemies of His Kingdom among men, it is to express thereby *the mind of God.*

described as "they that are with Him (the Lamb), called and chosen and faithful."

"And out of His mouth proceedeth a sharp sword, that with it He should smite the nations. And He shall rule them with a rod of iron. And He treadeth the winepress of the wrath of Almighty God." There is here a combination of two kinds of power, the sword of which Christ wields. The sword of power to gain (by the truth brought home to them) a *willing* people in the day of His power" (Ps. cx. 3).* These are the volunteers in this war, on Christ's side—"They that are with Him, the called and chosen and faithful." But there is also the sword of power to *subdue* the determined enemies of the truth, "with a rod of iron." Probably it is to such that the words of Ps. xlv. 3—5 are meant to apply, though they may be delightfully applied to the former class :—

"Gird thy sword upon thy thigh, O mighty
   one,

---

* "Thy people offer themselves willingly in the day of Thy
   power:
In the beauties of holiness, from the womb of the morning,
Thou hast the dew of Thy youth." (R. V.)

Thy glory and thy majesty.
And in thy majesty ride on prosperously,
Because of truth and meekness and righteousness :
And thy right hand shall teach thee terrible things.
Thine arrows are sharp ;
The peoples fall under thee ;
They are in the heart of the king's enemies."

"And He hath on His garment and on His thigh a name written, KING OF KINGS AND LORD OF LORDS." Hence verse 11, "upon His head are many diadems."

Now for the issue. It is not told in an impassioned prose, but in the stately style of this prophecy throughout :—

"And I saw an angel standing in the sun (in the gaze of the whole earth); and he cried with a loud voice, saying to all the fowls that fly in mid heaven, Come and be gathered together unto the great supper of God; that ye may eat the flesh of kings, and the flesh of captains, and the flesh of mighty men, and the flesh of horses and of them that sit on them, and the flesh of all men, both free and bond, and small and great. And I saw the beast, and the kings of the earth, and their armies, gathered together to make war against Him that sat upon the horse, and against His army. And the beast was taken, and with him the false prophet that wrought the signs in his sight, wherewith he deceived them that had received the mark of the beast, and them that worshipped his image: they twain were cast alive into the lake of fire that burneth with brimstone; and the rest were killed with the sword of Him that sat on the horse, even the sword which came forth out of His mouth.* And all the fowls were filled with their flesh."

---

* These—"the rest"—probably mean the hangers-on, who take no active part in the struggle, and perhaps little understand the great quarrel, but, being found on the wrong side, and an end made of their cause, are made to feel to their cost that that cause was indeed not the cause of God.

THE THOUSAND YEARS.

Chap. xx. 1—10.

SATAN BOUND.

"And I saw an angel coming down out of heaven, having the key of the abyss (Gr. 'bottomless deep') and a great chain in his hand. And he laid hold on the dragon, the old serpent, which is the Devil and Satan, and bound him for a thousand years, and cast him into the abyss, and sealed it over him, that he should deceive the nations no more, until the thousand years should be fulfilled" (xx 1—3).

So great diversity of opinion exists about every feature of this famous period, that I will try to do justice to the language of every clause of the text.

1. Is the "thousand years" to be taken as a definite for an indefinite period? I answer, Certainly not: (1) because the contrary is

clearly implied in the announcement that his confinement is limited to a *fixed period*, at the close of which he is to be set at liberty; and (2) because this period is expressed, not at all *symbolically*, like that other famous period of the unbroken reign of Antichrist in three different ways—"a thousand two hundred and sixty days"; "forty and two months"; "a time, and times, and half a time"—but nakedly in "years."

2. What are we to understand by the "deception" which the enemy is shut up from practising for a thousand years? Does it mean that there will be a total cessation of Satanic influence upon men for a thousand years? I answer, Certainly not. For (1) it would be against the whole tenor of Scripture teaching elsewhere; and it is in the last degree improbable that such a fact would be disclosed only in a symbolical book of much difficulty. (2) Since the purpose for which he is to be confined is "that he should deceive *the nations* no more" during that period—in other words, that he should be able to create no *public interest*, no *hostile cause*, to disturb the peaceful reign of Christ. The stage had been succes-

sively cleared of *Babylon*, *Persia*, *Greece*, and at length of *Rome* (Pagan and Papal); and there was to be an end of this struggle, even though terminating always in the defeat of the enemy. This is clearly the sense of the prediction. And if after the thousand years let loose again "for a little time," it is only to reveal his unchanged enmity to Him whose love, once forfeited, he can never regain, and therefore will do his worst to the last.

3. In what sense are we to understand the seizure of the dragon, the chaining of him, casting him down into the abyss and sealing it over him? Is it by an immediate putting forth of the Divine Hand to arrest his action against Christianity? or is it a victory to be gained by the heroic and united action of believing men against the enemies of their Lord—*truth* in them victorious over error? This may seem too weak an interpretation of the strong symbolism employed. But is not the same style of symbolism employed in previous chapters in predicting the overthrow of Rome Papal? In view of this I ventured—in a book on the Second Advent, published (1st edition) nearly fifty years

ago *—to express my belief that what is here predicted of the shutting up of the dragon would be effected, not by some exercise of sovereign power to *remove the enemy from the scene of conflict*, on which he had been defeated already, and so saving the Church from the necessity of fighting him with its own "whole armour of God"; but that such would be the victorious power of faith in the Church during that happy period that, the cause of Christ carrying it everywhere, none would be found to stand up for Satan, his trade would be at an end; and none appearing for him, it would be as if one should go in search of him, and find him nowhere.

But now, at this late date, and looking at the subject with a fresh eye, I am inclined to think that, over and above all I said before, a number of influences may co-operate to effect this *capture* of the old serpent. The whole state of society may be so changed as to afford no base of operations for the enemy to work upon; while, on the part of the Church, everything

---

* "Christ's Second Coming: Will it be Premillennial?" Seventh edition. T. & T. Clark, Edinburgh.

may be in its favour. And here, I think, we may safely leave the matter.

THE THOUSAND YEARS' REIGN WITH CHRIST.

Chap. xx. 4.

"And I saw thrones, and they sat upon them (or, they were sat upon): and judgment was given unto them (that sat upon them). And I saw the souls of them that had been beheaded for the testimony of Jesus, and for the word of God, and such as worshipped not the beast, neither his image, and received not the mark upon their forehead and upon their hand; and they lived, and reigned with Christ a thousand years."

This has been the great field of controversy between two sections of Christians from the third century to the present day. Up to the third century the persecuted Christians seem to have taken this as predicting a literal resurrection of the martyrs to reign with Christ. Such as demurred to this were called "allegorists"; but there is no evidence that they formed a party, or had any open controversy with the literalists, until, in the third century, Dionysius of Alexandria, a disciple of Origen, went down to Egypt to discuss the question with the literalists or Chiliasts ("millenarians"), who were warm advocates of the literal sense.

In modern times interpreters who reject the literal sense seem to me to argue the question on a false issue—I mean a misapprehension of what the text says. It was not the *souls* that were seen that lived and reigned; for the souls were not slain. It was the *martyrs themselves* that were seen living again, and reigning with Christ. And the one question is, Is the prediction to be understood *literally* or *figuratively*?

Well, plainly, whether literal or figurative, it is the *martyrs only* that are the subjects of this prediction. Had the language been quite general, there might have been room for some latitude as to the party intended. But it is quite the reverse. It is impossible for language to be more definite. They are two distinct sets of martyrs.

1. "Those who had been *beheaded* for the testimony of Jesus and for the word of God." Now "beheading" was a Roman mode of execution—as we see in the case of John the

Baptist (Mark vi. 27, 28); and this word seems used here expressly to intimate the *authority* by which the martyrs here meant were put to death. Besides, the *cause* of their death is expressed in the very terms used, by which the writer of this book tells us why he was in the isle of Patmos (i. 9), under a *Pagan* persecution. And still more, in chap. vi. 9—11, we have the very martyrs here described, crying from under the altar for "judgment to be given them" against those who slew them. The answer there given them was that they should have what they righteously claimed, but not until a second crop of martyrs, to succeed them, should be reaped; and this is the class that next comes before us in this vision,—described even more unmistakably than the first ones.

2. "Those who worshipped not the beast nor his image, nor had received the mark on their forehead and in their hands."

None who have read the preceding chapters of this book can mistake who are meant here; for by referring us back to chap. xiii. 4, 14—16, and xix. 20, they tell us plainly that they are that class of martyrs which was to succeed those under *Paganism*: in other words, the victims of the *Papal antichrist.* Of both these classes it is here said that "judgment was given unto them"—as promised in vi. 11. And what is the "judgment" they get? "They live, and reign with Christ a thousand years"— not surely all the saints that have lived up to that time, else why is this not *said*? If you will have literal interpretation, you must go through with it.

I have said that the early Christians who took the prediction literally understood it of the martyrs exclusively; insomuch that it begot a fanatical desire of martyrdom, that they "might have part in the first resurrection." Even some of the modern advocates of the literal sense have felt themselves obliged to admit that it will be that of the martyrs only. Others—feeling, no doubt, that a thousand years' reign with Christ of martyrs only is not likely to be the thing intended—are fain to include such as have been virtual martyrs, such as eminent confessors (so Bp. Newton). But without enlarging on this here, I must refer the reader to my book on the "Second Coming," where (in chap. ix.) he will find, in fuller detail than he is likely to find elsewhere, the various opinions of literalists on this point.

But at the present day I know of no interpreter of the Literalist School who does not extend " the first resurrection to *all the saints* " at His coming."* And what ought to surprise us is, that they seem to take this for granted, reasoning out the whole plan of the thousand years' reign, as if the " first resurrection " of all believers were the obvious sense of the prediction. But this will not go down with

* In his masterly "Commentary on the First Epistle to the Corinthians" (1885), the Rev. Principal Edwards, when commenting on chap. xv. 26, writes with less than his usual cautious accuracy. After saying: "It seems, it must be confessed, that the apostle teaches that there will be two resurrections—the former of believers only; the latter of all others, when at last 'death itself shall die:' the first resurrection is the redemption of the body, for which all believers groan (Rom. viii. 23),"—Dr. Edwards adds: "Similarly the apostle John says that the followers of Jesus—not the martyrs only; for καὶ οἵ τινες ('and such as') introduces others—will rise and reign with Christ a thousand years. And this is the first resurrection. But the rest of the dead will not rise till the thousand years are ended" (pp. 417-18). That the saints will rise before the wicked is beyond doubt true. But the apostle John does not say that "the followers of Jesus (not the martyrs only) will rise and reign with Christ a thousand years." It is Dr. Edwards only who makes him say that. The apostle says that those who were *beheaded* will rise; and the next clause does indeed "introduce others," but it is *another class of martyrs*, it is *martyrs only*, of whom Rev. xx. 3, 4, speaks, and I must insist on not loose but exact interpretation.

exact interpreters, who will insist that if required to believe the first resurrection is to be a literal one, the *party* that is to rise must be taken with the same literality.

I admit that when one reads the two verses that follow, the literal sense of them seems more natural than the figurative. But since this is confessedly a symbolical book, ought not its own symbolical predictions elsewhere to guide us in the interpretation of this one? Can anything express more literally the killing of the two witnesses (xi.), the exposure of their dead bodies for three days and a half in the streets of the great city, their resurrection after that, and ascension into heaven? (See the remarks on this in the preceding pages.)

But why go about to prove from Scripture that *life from the dead* is familiar to us in every form of the idea? Is it not current coin in every language and every age? A striking illustration of this way of speaking, when no literal resurrection of the dead is meant, is recorded of John Huss, when in prison at Constance awaiting his execution :—

" When the venerable priest (says Merle d'Aubigné) had been summoned by Sigismund's (the Emperor's) order before the Council of Constance, and had been thrown into prison, the chapel of Bethlehem, in which he had proclaimed the Gospel and the future triumphs of Christ, occupied his mind much more than his own defence. One night the holy martyr saw in imagination from the depths of his dungeon the pictures of Christ that he had painted on the walls of his oratory effaced by the Pope and his bishops. This vision distressed him; but on the next day he saw many painters occupied in restoring these figures in greater number and in brighter colours. As soon as their task was ended, the painters, who were surrounded by an immense crowd, exclaimed, 'Now let the Pope and bishops come, they will never efface them more.' 'And many people rejoiced in Bethlehem, and I with them,' said John Huss. 'Busy yourself with your defence, rather than with your dreams,' said his faithful friend the Knight of Chlum, to whom he had communicated the vision. 'I am no dreamer,' said Huss, 'but I maintain that the image of Christ will never be effaced, but it will be painted afresh in all hearts by much better painters than myself. The nation that loves Christ will rejoice in this; and I, AWAKING FROM AMONG THE DEAD, AND RISING, AS IT WERE FROM THE GRAVE, SHALL LIVE WITH GREAT JOY.'"*

But, after all, I confess that, with me, the conclusive argument against the literal sense of this prediction—viewing it as the resurrection

---

* "History of the Reformation in the Sixteenth Century," vol. i., p. 23. (Oliver & Boyd: Edinburgh, 1846.)

of all the saints that have lived before the second coming of Christ, and the thousand years' reign to succeed that—is the theory of *what the state of things will then be,* which we are asked to accept.

1. "The dead in Christ will rise first, and all in Christ that are found alive at His coming ("their mortality being swallowed up in life") will be caught up together with them in the clouds, to meet the Lord in the air, and so shall they ever be with the Lord" (1 Thess. iv. 16, 17). This is *heaven*—the state of glory—the saints, there with Christ, in glorified bodies.

Modifications of this (as the secret rapture and the leaving behind of the unready for a limited period) I touch not, confining myself to the essentials of the theory.

2. There will then be left on earth a whole human race, in mortal flesh and blood, as now, for a thousand years. What will be their condition? They will have to be made Christians—that is to say, believing in God and in Jesus Christ. But what their faith and hope will rest upon it is impossible to conceive. For if *our* Scripture is their only revelation,

what can *they* get out of it? Christ's Second Coming is the hope of the Church now, but after that is past there is no third Coming that they can look for. Nor can they tell what will become of them when they die. They have no Scripture to assure them that they will rise again; nor if they do, in what state and to what futurity. For all that relates to saints in Scripture has to do only with those who have lived *before* the Second Advent. Accordingly, some (as the late excellent Edward Bickersteth) go the length of believing that the earth will be peopled by a race of men in mortal flesh and blood, marrying and giving in marriage, as now, *for ever*. Others, not prepared for that, suppose the saints on earth after Christ's coming will have a resurrection from the dead, and to life everlasting, but on a lower plane than that of the glory of the heavenly state, and not "with Christ" (probably) so immediately.

But what is all this, but *mere speculation*, based upon nothing in Scripture where the foot can stand? So obvious is this, that some expect *a new revelation* for their behoof. But this also is pure conjecture.

Now, what answer do we get to all this? Just no answer at all—except that these are things which we have no need to trouble ourselves about; that "the day will declare it." Well, there are many things about the future on which we should desire light, but on which Scripture gives none—such as whether and to what extent there will be a recognition of each other by souls in heaven, of what nature the "many mansions" will be, whether there will be diversified occupations as well as different degrees of nearness to the throne; also whether Christ will be seen in human form, and as such will hold fellowship with them. On all such questions Scripture is silent, and it is good for us to be content to wait for light on them till the time come.

But what the literal "first resurrection," as now understood in connection with the pre-millennial theory of Christ's Second Coming, leaves *absolutely in the dark*, is of a totally different nature, and *we refuse to be put off*. When you tell us of a human race to people the earth for a thousand years after Christ at His Second Coming has taken up with Him to glory all the saints that have lived before that time, we have a right to demand some answer to such questions as I have named, about the race you speak of,—how they are to be converted, what is to be their faith and hope—in fact, whether their SAINTSHIP will be the same as ours, and what their hopes for the future. If all the answer they have to give is to give it the go-by, I hold it fatal to the whole theory. In short, for myself, I must set aside my New Testament before I can believe in it. I said this fifty years ago; and it is to me more incredible now, if that be possible, than it was then.

But if "the first resurrection" is not to be a literal one, how are we to conceive of it figuratively? The answer to this question is to be found in the fact that it is the *martyrs only* who are said to rise—those under the *Pagan* and the *Papal* persecutions. And just as it is said of Abel, that his blood cried unto God from the ground, and that "he being dead yet speaketh"; and just as John Huss was certain that his testimony for Christ, though slain in his person, would rise to life in others who would be "better preachers than he," so the witness borne by that noble army of martyrs, for

Christianity against Paganism, and for Christ against Antichrist, would rise and triumph and be no more disturbed for a thousand years.

To expatiate here on what will then be witnessed would be unsuitable in this A B C. In my book on the Second Advent I did my best to point out, in the light of Scripture, what would be its leading features (chap. viii., Part II.), and to this I refer the reader. One thing, however, is clear: that "blessed and holy" must he be " who shall have part in this first resurrection," and " over them " it is certain that " the second death shall have no power "—for that is a death that will never die—but " they shall be priests of God and of Christ, and shall reign with Him a thousand years."

THE REST OF THE DEAD.

" The rest of the dead lived not again till the thousand years should be finished " (ver. 5). But what happened *then*? Did they "live again " (*literally* on the thousand years being finished) ? Not a word of this do we read. But if the sense is *figurative*, we see them here alive, and ready tools of the serpent when again let loose. He was shut up that he might deceive *the nations* no more for a thousand years. But, this long period ended, he is "loosed out of his prison," and finds "nations" enough prepared to do his old work. This is to me another clear proof that the whole language of the prediction is to be understood in a figurative sense.

"And when the thousand years are finished Satan shall be loosed out of his prison, and shall come forth to deceive the nations which are in the four corners of the earth, Gog and Magog, to gather them together to the war, the number of whom is as the sand of the sea" (ver. 8). Let us look at this statement first in the *figurative* and then in the *literal* sense.

To Christians whose lot is cast within the bright latter day, it must be a joy to live,—religion is uppermost everywhere and in everything,—in public affairs, business transactions, in social intercourse, in family life; worldliness quite out of fashion, and the ungodly nowhere to be seen, or obliged to "feign submission": a "heaven upon earth," and this for a thousand years!

"But the sun of this bright day is destined to set—*gradually*, no doubt, as gradually it will rise, subject to the laws of spiritual decline. . . . Everything may go on as before, but without the 'first love,' with the mechanism of habit, and lingering recollections of the past. By this the jealous Lord of the Church will be touched, and His Spirit be grieved, whose withdrawal will accelerate the decline. . . . Settling upon her lees, and her outward prosperity proving a snare to her, secularity taking the place of spirituality, and inconsistencies appearing increasingly, the Church's influence for good upon the world at large will grow less and less. And just as on a small scale in some little community like *Northampton*, as described by President Edwards,* after the remarkable sense of God's presence over the whole town had begun to wax feeble, the still unconverted portion of it, though subdued and seemingly won over to Christ, would by little and little recover themselves, and at length venture forth in their true character,—so will it probably be on a vast scale at the close of the latter day. The unconverted portion of the world, long constrained by the religious influences surrounding them, to fall in with the spirit of the day, catching some of its wholesome impulses, but never coming savingly under its power—this portion of mankind, which we have reason to believe will not be small, will now be freed from these irksome restraints, no longer obliged to breathe an atmosphere uncongenial to their nature, and 'feign

---

* See "Second Advent," p. 404.

submission.' . . . And then the Lord will be provoked to let loose upon them the roaring lion." *

Such is an intelligible, and, I humbly think, a reasonable view of the prediction, when taken in a figurative sense. But now try the literal sense of it, and I hesitate not to say it is absolutely unintelligible. For where did those myriads of " nations " come from, whom Satan, as soon as he is "loosed from his prison," finds all ready to be "deceived,"—that cloud of locusts to eat up the Church of Christ? Since all the saints that have lived before the thousand years and the Second Coming of Christ, are supposed to have risen from the dead to reign with Christ, and "the rest of the dead lived not again till the thousand years were finished," did they rise from the dead *then*? It is not *said* that they did, but we must presume that they did so, if we are to interpret literally. Is it *they*, then, that are the "nations" whom Satan "gathered together against Christ"—men risen from the dead, in resurrection-bodies, to war on earth against the Church? *Will this be credited by anybody?* But if not, *the literal sense will not do.*

---

* "Christ's Second Coming," etc., pp. 414-15.

SATAN'S LAST EFFORT AND FINAL DEFEAT.

"And they went up over the breadth of the earth, and they compassed the camp of the saints round about and the beloved city" (ver. 9). The Church is supposed to be encamped in the wilderness, and "the beloved city"—the city of the great King, their capital. And the names given to the "nations" that are "gathered" by Satan against Christ, "Gog and Magog," carry us back to that unprovoked, formidable, but abortive attack upon the people of Israel, after they were re-settled in their own land (Ezek. xxxviii., xxxix.).

Whatever this may mean, it is not to be supposed that it will begin and end all at once. "The little time" during which Satan is "loosed from his prison" must be read as "little" relatively to the thousand years of his imprisonment; and so it may extend through a century or two, for aught we can tell. "Since it cannot be imagined (says *Faber**) that the whole world will plunge at once from piety into impiety, both common sense and general expe-

---

* "Sacred Calendar of Prophecy," iii., p. 478.

rience may teach us that a considerable time will elapse ere the children of men will become so thoroughly depraved as to enter into a regular combination for the purpose of extirpating the small remnant of God's faithful people." The names "Gog and Magog," to express "the nations" that thus combine, refer back to those who after the re-settlement of Israel securely in their own land, made a formidable but abortive attack upon them (Ezek. xxxviii., xxxix.).

Thus surrounded by overpowering numbers, ready to burst in upon and make an end of them, one naturally asks what was to be done by the faithful? I suppose something like what faithful Hezekiah and his people did, when Jerusalem was invested by an army sufficient to overwhelm them,—they lifted up their prayer for the remnant that was left, saying, " This day is a day of trouble, and of rebuke, and of blasphemy; for the children are come to the birth, and there is not strength to bring forth. . . . Incline thine ear, O Lord, and hear; open thine eyes, O Lord, and see; and hear the words of Rabshakeh, which hath sent to reproach the living God" (Isa. xxxvii. 3, 17).

Their tremulous cries enter into the ears of the Lord of Sabaoth. "And shall not God avenge His own elect, who cry unto Him day and night? I say unto you, He will avenge them speedily" (Luke xvii. 7, 8). And just as "when the sun rose upon the earth when Lot entered Zoar, the Lord rained upon Sodom and upon Gomorrha brimstone and fire from the Lord out of heaven" (Gen. xix. 23, 24), so when the final attack upon the Church of Christ and cause of God seems ready to crush it out of existence, "fire came down out of heaven and devoured them. And the devil that deceived them was cast into the lake of fire and brimstone, where are also the beast and the false prophet: and they shall be tormented day and night for ever and ever" (vers. 9, 10).

THE GENERAL JUDGMENT.

Chap. xx. 11—15.

"And I saw a great white throne, and Him that sat upon it, from Whose face the earth and the heaven fled away, and there was found no place for them. And I saw the dead, the great and the small, standing before the throne: and books were opened, and another book was opened, which is the book of life. And the dead were judged out of those things which were written in the books, according to their works. And the sea gave up the dead which were in it, and death and Hades gave up the dead which were in them. And they were judged, every man according to their works. And death and Hades were cast into the lake of fire. This is the second death, even the lake of fire. And if any was not found written in the book of life, he was cast into the lake of fire."

"If ever language could express the doctrine of a *simultaneous and universal* resurrection of the dead, surely we have it here. Who that credits what is here stated could ever imagine that all mankind were not in this august scene, in their resurrection state, and that himself would not form part of it? But premillennialists see none but *the wicked* here, and even of them only such as have died before the millennium."* The early premillennialists never dreamt of such a thing; but the necessities of the theory, as now held, rendered it certain it must come to this; and hard work they have of it to make their interpretation in the least plausible.† The whole scene so speaks for itself that I will only call attention to its principal features.

The scene opens with surpassing majesty. The great throne is "white" in the spotless purity of the Judge and the unerring rectitude of His judgments. The lustre of His face is so appalling that the earth and the heaven at the sight of it fled out of existence! But the dead, small and great, have to face it; for they are there each one to be judged according to his works. For this purpose, "*books* were opened, and the dead were judged out of those things which were written in the books, according to their works." This done, and each case already decided, "another book was opened, which is the book of life." We have seen already what this "other book" is. It is a book not of "*works*"; for these are contained in "the books" which have been already produced and told their tale. It is a book of *names*—

---

* "Second Coming," etc. (7th ed.), p. 195.
† See the extracts in the place just quoted.

"written from the foundation of the world in the book of life of the Lamb that hath been slain" (xiii. 8; xvii. 8). For what purpose is this "other book" opened? Doubtless to shew that "known unto God are all His works from the beginning of the world" (Acts xv. 18); and that "whom He foreordaineth them He also calleth, and whom He calleth them He also justifieth, and whom He justifieth them He also glorifieth" (Rom. viii. 29); yet that each one of these has to pass the judgment according to his works, and that if he is *adjudged* to life, it is not because his name has been written from the foundation of the world in the *book* of life, but because "his *works* do follow him." The "other book" simply *countersigns* the judgment already passed, "according to the things that were written in the books."

THE LAST THINGS.

CHAPS. xxi., xxii.

It is not within the scope of my plan, in this A B C of the Apocalypse, to take up in detail these closing chapters. That they *follow*, in point of time, the Last Judgment (chap. xx.), is to me quite clear. It is true that we read (chap. xxi. 24, 26, and xxii. 2), that "the nations" shall walk in (or "by") the light of "the New Jerusalem," and that "the kings of the earth do bring their glory into it"—as if they themselves had no part in it; also, that the leaves of the "tree of life were for the healing of the nations"—implying, apparently, that after the redeemed have entered into glory there will be in "the new heavens and the new earth," "nations" *as such*, with their "*kings*," still in flesh and blood. Accordingly, those who believe that the Second Advent will be premillennial, hold this to be conclusive evidence that the two last chapters of the Apocalypse belong to the Thousand Years. In that case, it follows that *the final condition of the human race is not revealed at all.* But few devout and unprejudiced students of the New Testament will believe this. I admit the difficulty of otherwise explaining the verses which I have quoted. But the difficulties of the other theory are immensely greater, and (as I have shown elsewhere) involve what is self-contradictory.

On a survey of the contents of this volume, one or two concluding remarks may not be unacceptable to its readers.

1. What I believe I have made good by strict exegesis of the text, I have expressed confidently. (1) If in the Apocalypse there is one *fixed quantity* (to speak algebraically)—one prediction on which we can plant our foot as sure ground—it is that of chap. vi. 10, 11—that there were to be two *protracted periods of cruel martyrdom* of the saints and faithful in Christ Jesus,—the one by *Pagan Rome*, the other, in succession to it, by *Papal Rome*; also, that in chap. xx. 4, *both these companies of martyrs were beheld* by the Seer *living again and reigning with Christ for a thousand years*—in the triumph of what they died for, in their successors. But there are not a few details—some of them of much importance—of which I could only give what appears to me to be the thing intended. This distinction I hope will be borne in mind. Indeed, some details I do not see my way to at all, and therefore I meddle not with them. The most important of these

details are those of the *Trumpet Series*, chap. viii. 7—ix. Two things about this Series I spoke confidently about: that they were a succession of *judgments on some interest hostile to Christ*; and that this interest was *Papal Christendom*. The evidence for this the reader will find in pp. 104-126.* But the details of each successive judgment, in this Trumpet-Septenary, I leave to other and better interpreters.

2. In no other book of the Bible do we find such threatenings against any one who shall dare to tamper with its contents, by either

---

* In confirmation of this, I here add what I should there have begun with. These judgments were brought down upon the hostile interest *in answer to the prayers of all the saints*. For the angel who offered these prayers—" adding much incense " to them, as the symbol of how acceptable they were to the Hearer of prayer (compare Ps. cxli. 2)—took the censer, from which the incense ascended, and filled it with the fire of the altar, and cast it upon the earth, and there followed thunder and voices and lightnings and an earthquake " (viii. 3—5), whereupon the judgments, one by one, are announced, to the last. And what are these "prayers of all the saints"? They correspond to the "cry" from under the altar of the souls that had been slain under the *Pagan* persecutions. They asked how long would their " Master " delay to "avenge their blood on them that dwell on the earth"? and they were told they would have to wait till another crop of martyrs, to succeed them, should be reaped (chap. vi. 10, 11). And these "prayers of all the saints" are the cry of the martyrs of *Papal Christendom* for the avenging of *their* blood.

adding to or taking away from "the things that are written in this book." To him who shall *add* to them, "God will add the plagues written in this book." From him who *takes away* from them "God will take his part out of the book of life, and out of the holy city, and of the things that are written in this book."* Whatever this may import, one thing at least may be said of it: that to neglect altogether "the prophecy of this book"—to take no pains to understand what it was meant to teach—is not the way to avert this threatening; much less to make a boast of passing it by as a hopeless study, and look down upon those who waste their time upon it.

3. It will surprise many, perhaps most readers of the Apocalypse, to be told that, though it is "the book of the revelation of Jesus Christ," His *Second Coming*—as to either the *time* or the *manner* of it—*is not within the limits of its predicted future:* I mean the strictly *prophetical* part of the book (from chap. iv. to xxii. 6 inclusive). In what goes before this

---

* I advisedly adhere to the received text here and the A.V., as against the R.V.—both on internal grounds and the textual evidence. But I cannot stay to make this good.

it comes out abundantly. Even in the Salutation the Seer breaks forth in rapturous strains, "Behold, He cometh with the clouds, and every eye shall see Him." To the church of Thyatira the glorified Head of all the churches says, "That which ye have, hold fast till I come." In fact, every promise "to him that overcometh" in the seven churches points to and presupposes His Second Coming as the time when they will be all realised.

So much for what *precedes* the strictly prophetic part of the book. But in the strictly *prophetic* part of the book (chap. iv.—xxii. 6) not a word is said even about the fact, that He *will* come, nor any indication of *when*—relatively to the other events there predicted—His coming will take place. To point to the words in chap. xvi. 15, "Behold I come as a thief: blessed is he that watcheth," proves nothing; for over and over again these warnings are given to prepare for portentous events manifestly short of His Second Coming, and in each case the *connexion* must determine the event referred to. That chap. xix. 11, etc., announces the Second Coming of Christ will hardly be affirmed by those who observe what is said to follow the events there predicted.

What I say, then, is, that the bright hope that He " will appear the Second time, without sin, unto the salvation of them that look for Him " is *not once even alluded to within the strictly predictive part of this " book of the revelation of Jesus Christ."* But no sooner does this end, in chap. xxii. 6, than the Master Himself takes speech in hand, and in verse 7 —as if impatient to get the reader out of all that sort of thing—breaks forth, " Behold, I come quickly : Blessed is he that keepeth the words of the prophecy of this book" (now completed). All that follows is supplementary—a rich *winding-up* of a book which, *sui generis*, is incomparable. But in the heart of it again He says, " Behold, I come quickly"; and at the close, " He which testifieth these things saith, Yea, I come quickly. Amen." " Yea, come, Lord Jesus," is the glad response of the writer, and of all who love His name !

Now, what do we gather from this *studious* absence from all the *chronological* predictions of this book, of any allusion to what is of infinitely greater importance to every Christian than any of the predictions which have been the chief subject of this volume—the Second Coming of Him whose words are graven upon His heart, and almost ever on His lips,—" I go to prepare a place for you, and if I go and prepare a place for you, I will come again and receive you unto Myself, that where I am, there ye may be also"? It is because *this Coming is of a perfectly different character from all "comings," either "quickly" or "as a thief" in the strictly prophetic part of this book*—not to say in the prophecy of our Lord in Matt. xxiv. and Luke xix., in which His "*coming* with the clouds of heaven with power and great glory" is studiously expressed in the *language* of His Personal Advent; though Matt. xxiv. 34 makes it perfectly clear that the event predicted was the destruction of Jerusalem, the breaking up of the Jewish State, and the dispersion of the Jews among all nations. And the reason is plain. *Nations* or *public bodies can only be judged and sentenced to perish in the present world.* But when they are so, it is their *Judgment Day*; and it is fitting it should be expressed *as such*, in order that we, in these later days, should read, *in that phraseology*, what will be to us "the judgment of the great day."

But let any one read how the Second Personal Coming of Christ is expressed in those portions of the New Testament which are of a *non-predictive* character, and he will at once recognise the difference I wish to point out and insist on.

Yes, that bright day which has been the hope of the Church in every age, is not a question of *figures*, but of *faith*. It was "at hand" in the apostles' day; it is at hand to-day, and *in the same sense*; and if He should not come for a thousand years more, it will still be "at hand," and in the very same sense. Chronologically, this was not true in the apostles' day, nor for more than a millennium after. But no matter. He will come, says the expectant Christian, *when He is ready to come, and I am ready when He is*. And faith *sees* Him coming, "leaping upon the mountains and skipping upon the hills." And neither in the spirit of sloth and carnality, does it say, " My Lord delayeth His coming," nor in the spirit of fanatical and excited expectation about a present appearance, but in that sublime attitude which the apostle calls " the patience of hope," it is the privilege of faith to say—alike when chronologically near, as in holy defiance of mere dates, because ready for them all alike—" Make haste, my Beloved, and be Thou like to a roe, or to a young hart upon the mountains of spices " (Cant. viii. 14).*

---

* "Christ's Second Coming," etc. (7th Ed.), page 51.

ADDENDA.

I.

[When commenting on chap. xii. 6, 14, I reserved what I had to say about the question *Where* is this "wilderness" to which "the woman fled from the face of the serpent"? But not being satisfied as to the proper place for inserting what I wrote on this point, it escaped my memory altogether. And as the question is one of some interest, I supply it here.]

BUT *where* is this "wilderness"? it will be asked. Most interpreters reply, In the valleys of the south of France and Savoy, *Dauphiny*, *Provence*, and the High Alps. And thither, beyond doubt, a large body of the faithful fled for refuge from the bloodhounds of *the Church*. There for a long time they were left undisturbed. For an utter indifference to religion reigned there, amid the luxury and laxity of morals in which the court and wealthy nobility spent their lives; while the people were equally lax in their own—neglected by the clergy, whose own lives were a disgrace to religion. Taking advantage of this, the refugees began quietly to insinuate into the families of the people those saving truths for which they had had to fly. Some went among them as preachers, but others with more success as pack merchants, selling jewellery and other such articles, but with New Testaments and religious books secreted in their baggage. Having once gained the ear of a family, and awakened their curiosity—by telling them of priceless jewels which they possessed but could not sell—they would be asked to show them, and after some hesitation would cautiously open their treasures, and, refusing all offers to purchase them, agree to give them for nothing, on a promise not to betray them. In this way converts of whole families were rapidly made, until the whole region seemed to swarm with "heretics."

That this is no exaggeration we have the best of all evidence, from one who had been himself a leading man among the Waldenses—*Reinerius Sacho* or *Sacco*—but who turned his back upon them, became a Dominican monk, and then an *Inquisitor*, and one of the earliest after that inhuman tribunal was established (in 1229). This man published a book, in the course of which he describes, very much as I have done, the way in which those *colporteur* preachers stole away the hearts of the people.*

Naturally enough, having no regular ministry, these simple men would make mistakes in their teaching; and these being exaggerated,† the mistakes of a few would be ascribed to the whole body. So a hue and cry of "heresy" was got up against them. Had their moral character been as bad as it was the reverse, they would never have been troubled. But a word against the teaching of the Church was not to be tolerated. And Innocent III. had scarcely succeeded to the throne of the Church, in 1198, when that haughty Pontiff thought to signalise his reign by proceedings against this sect. But he was wise in his generation, and having learnt how extensive was the defection, and unblemished their character, he made offers to them of concessions with a view to conciliate

---

* Curiously enough, though I never saw the book itself, in a box of books which I had bought in London about fifty years ago, when residing there, from the library of a retired Indian, and afterwards sent me to Scotland, I found the interstices of the box stuffed with some twenty or thirty leaves of an old Latin book, and on reading them I could easily identify them with our Inquisitor's book, for they told the very tale of the pack-merchant preachers—which I wish I had not destroyed.

† But, as a friend once said to me of a similar case, for one error they would teach they would teach twenty truths, and the one correct the other.

them. But it was too late. And, chafed at their refusal to return to the bosom of the Church, he set on foot, in 1209, "an atrocious series of crusades, led by Simon de Montfort, Earl of Leicester, and extending over thirty years. By this terrific war the swarming misbelievers of Provence were almost literally drowned in blood. The remnant which escaped the sword of the Crusaders fell a prey to ruthless agents of the Inquisition—the Tribunal now established permanently by the Council of Toulouse, 1229, for noting and extinguishing all kinds of heretical pravity."*

Mezerai ("Hist. de France") † tells us that the pretext for this war was an accusation brought against Raymond, Count of Toulouse, of harbouring the Albigenses in his dominions, and that an army of 400,000 cross-bearers was raised against him, among whom were five or six bishops. They took the town of Beziers, and put all to the sword, to the number of 60,000—pursuing the war with like cruelty and fury in many other places; and Montfort, the general of this holy war, was rewarded with the greatest part of the Count's dominions, while

---

* Hardwick's "History of the Christian Church" (Middle Age), p. 310 (1853).
† Quoted by Fraser, "On the Prophecies," p. 114.

the Count himself was deposed, as a favourer of heretics, Montfort being declared lord of all the countries he had conquered.*

Such is the answer for the most part given, to the question *Where* was that "wilderness" to which "the woman fled from the face of the serpent" (chap. xii. 6, 14)?

But I see no reason why we should restrict the "wilderness" to which "the woman fled from the face of the serpent" to the South of France, or *localise* it at all. Christ's faithful witnesses were not confined to France or any

---

* A large body of those faithful witnesses fled from their pursuers up the High Alps, and when still pursued, went higher and higher up, till their enemies, unable to face a region of almost perpetual snow, gave up the pursuit. In that inhospitable solitude the refugees settled down—making for themselves little hamlets along the line of a parish (as it afterwards became), eighty miles long. By degrees, having neither pastors nor teachers, the children lost nearly all that their fathers had suffered for, and at length almost their French tongue, speaking a mongrel *patois*. Nor did any one seem willing to encounter the rigour of that climate, as a missionary pastor, till Felix Neff—full of fruitful work elsewhere—got it made known to Louis Philippe, during his reign, that he would willingly exchange a paradisaical parish in the Valley for that of the High Alps, if he had an offer of it. And he had it; and the story of that heroic martyr to a climate to which he vainly hoped to naturalise himself, is told n a small 32mo book published by the London Religious Tract Society—which I regret to learn they have allowed to go out of print.

one country. Throughout all Papal Christendom "hidden" ones were to be found—little knots here and there, who sighed and cried over the hideous antichristianism called "the holy Catholic Church"—a gigantic State-Church, with a priesthood of boundless ambition, holding itself as much above the secular powers as the spiritual is above the corporeal and ought to rule it—a priesthood on whose altars sacrifice is offered for the sins both of the living and the dead, but which had no food for hungry souls. Well might one say "the word of the Lord was precious in those days: there was no open vision"! It was "not a famine of bread, nor a thirst for water, but of hearing the word of the Lord" (Amos viii. 11); "gleanings, as the shaking of an olive tree, two or three berries on the top of the uppermost bough, four or five in the outmost branches of a fruitful tree" (Isa. xvii. 6). In other words, the faithful witnesses of Christ were, during the whole of that protracted period of 1260 years of cruel persecution of all who dared to refuse its blasphemous claims, invisible—the true *Church of Christ* was *nowhere to be seen*; it wanted that "note" of a true Church, of which the Church of Rome boasts that it is possessed by it alone—VISIBILITY.

Returning for a moment to the Valleys, after the savage massacre of 1209, "heretics," we may well suppose, would be slow to settle there again. But we know well how, as late as 1655, the Duke of Savoy found so large a number of his subjects who refused to return to *the Church*, that, failing all missionary efforts to convert them, he determined to send an army against them, requiring them either to submit or at once leave the country, and, on their refusing both, to put them to the sword. This roused the indignation of Oliver Cromwell, who had by that time risen to supreme power, as Protector of the Commonwealth of England.

"The object of his foreign policy was to unite the Protestant State, with Britain at their head, in a defensive league against Popery, then and now the enemy of civil and religious liberty. Spain (the great under-propper of the Roman Babylon, the natural enemy of the honest interest) he determined to humble, and in due time he did. With France, less subject to the yoke of Rome, he allied himself, making such terms as he pleased; extorting from the Mazarin, a prince of the Church of Rome, protection for Rome's enemies, and full pardon for offences committed against her in the heart of France itself. In the summer of 1655, the persecution of the Protestants in the valleys of Piedmont afforded an occasion for displaying, in the noblest light, the greatness of the Protector, and of the nation which he represented. The tidings of these cruel oppressions affected the stern conqueror to tears. The treaty with France was ready to be signed that day. He refused to put his name to it until he received assurance of protection for the persecuted Piedmontese, and immediately wrote, not only to the Duke of Savoy himself, but to Louis XIV., to Cardinal Mazarin, the Kings of Sweden and Denmark, the States-General, the Swiss Cantons, and even Ragotaki, the Prince of Transylvania, pleading for their interposition. Had his remonstrances proved unsuccessful, he had fully prepared to exact compliance at the point of the sword. A protector not of the British realm only, but of the Protestantism of Europe, this 'usurper' might claim, without fiction, the title of 'Defender of the Faith.' Meantime, the supremacy of England on the seas was upheld by Blake, whose guns thundered on the shores of the Mediterranean, exacting justice and submission from every hostile power." *

---

* *Encycl. Britann.* (9th ed.), vol. vi., art. CROMWELL, p. 603. See also Carlyle's "Oliver Cromwell's Letters and Speeches," at the date referred to.

It was those frightful massacres of the Duke of Savoy's army, and the indignation which they roused throughout all Protestant Europe, that kindled the inspiration of our Milton to write that noble sonnet:—

> "Avenge, O Lord, Thy slaughtered saints, whose bones
> Lie scattered on the Alpine mountains cold;
> Even them who kept Thy truth so pure of old
> When all our fathers worshipped stones,
> Forget not: in Thy book record their groans
> Who were Thy sheep, and in their ancient fold
> Slain by the bloody Piedmontese that rolled
> Mother with infant down the rocks. Their moans

ADDENDA.

"*And I saw under the altar the souls of them that had been slain for the word of God, and for the testimony which they held: and they cried with a great voice, saying, How long, O Master, the holy and true, dost Thou not judge and avenge our blood on them that dwell on the earth? And there was given them, to each one, a white robe; and it was said unto them, that they should rest yet for a little time, until their fellow-servants also and their brethren, which should be killed even as they were, should be fulfilled*" (Rev. VI. 9-11).

II.

In quoting chap. xv. 6, I used the A.V. advisedly, and in a note (p. 128), I stated that the text adopted in the R.V., in one clause of this verse, was one of the worst in that version. I refer to the clause "clothed in pure and white linen"—for which the R.V. has "arrayed with [precious] stone, pure and bright," *marg.*

> The vales redoubled to the hills, and they
> To heaven. Their martyred blood and ashes sow
> O'er all the Italian fields, where still doth sway
> The triple tyrant; that from these may grow
> A hundredfold, who, having learned Thy ways,
> Early may fly the Babylonian woe."

"Many ancient authorities read *in linen*." That the "many ancient authorities" followed by the A.V. are right here, I showed at some length elsewhere; and for the sake of those who take an interest in textual criticism, I may be allowed to transfer to these pages what I wrote several years ago, in two articles, to illustrate what I take to be the fallacious principle on which the text of the Greek New Testament of Drs. Westcott and Hort is constructed—as in this case, and many others followed in the R.V. Dr. Hort has written a masterly volume, entitled "*Introduction* to the Greek Testament"; and as that Greek Testament is now much used in the Universities, it is but fair in this case at least *audire alteram partem*.

In the articles referred to I gave a number of examples of the *impossible* readings which the textual principle of Drs. Westcott and Hort's Greek Testament necessitated, to show its fallaciousness; and this being one of them, I set down here the criticism which I wrote upon it.

"Though I should be sorry to treat this reading in the bantering style of Sir Edmund Beckett, his arguments against it, which had occurred to myself before

I read them,* will not be easily refuted. But, as I have something of my own to say, I will put the case in the way that suits my own ideas. (1) Angels are often represented in Scripture as appearing to men, and usually their *dress*—evidently symbolical—is described. But where do we ever read of their having anything upon them of the nature of *ornament*? and does not the idea of a company of *ornamented angels* issuing from the temple strike one as something ludicrous? But (2) even though this difficulty could be got over, would it be natural, supposing this to be the true reading, to express the sense of it in this form—'clothed with stone,' in the singular number, and with no adjective to denote the quality of the stone? To obviate this manifest incongruity, the Revisers have inserted—with what right? the reader may ask—the supplement '*precious*,' and, still further to relieve the repulsiveness of the statement, have rendered it 'arrayed with *precious* stone.' But, in vindication of what Dr. Hort justly calls 'the bold image' of this reading, he refers, after Tischendorf, to the LXX. rendering of Ezek. xxviii. 13, as the Latin fathers who so read were fain to do. But the cases are entirely different. In the passage referred to the King of Tyrus in his pride is thus addressed:—'Thou sealest up the sum, full of wisdom, and perfect in beauty. Thou hast been in Eden, the garden of God; thou hast been clothed with every precious stone, the sardius, topaz, and the diamond, the beryl, the onyx, and the jasper, the sapphire, the emerald, and the carbuncle, and gold.'

---

* "Should the Revised New Testament be Authorised?" Murray, 1882), pp. 179-82.

To which I answer, it is one thing to describe, as beheld in vision, a proud *mortal man* strutting about in all the splendour that nine or ten of the most precious stones would make him to glitter in—after the manner of Oriental monarchs—a splendour of which he was soon to be stripped; but a very different thing, surely, it would be to describe a vision of *angels* issuing forth from the temple of God *so decorated*, for the purpose of pouring out vials of wrath upon an accursed system or its votaries. In the former case costly ornaments, such as Orientals pride themselves upon in proportion to their rarity and number, suitably represent the very thing intended: in the latter the incongruity is too apparent to need pointing out. Besides, in Ezekiel the stones are specified, and the word 'precious' is suitably added, even though their names might have rendered that superfluous—' every precious stone' ($\pi\acute{\alpha}\nu\tau\alpha$ $\lambda\acute{\iota}\theta o\nu$ $\chi\rho\eta\sigma\tau\acute{o}\nu$): whereas here the naked word "stone"—in the singular number, and with no qualifying adjective—is enough to condemn the reading as plainly an error, and all the rather as it is created by the mere substitution of one letter for another—the two words being otherwise identical. (3) Though the epithets 'pure *and* bright' are applicable enough to precious stones, it is worthy of note that in another chapter of this same book these are the words which are applied to the 'fine *linen*' in which the bride, the Lamb's wife, was seen arrayed (Rev. xix. 8). To this it is replied that there the word used for 'fine linen' is its usual word, $\beta\acute{\upsilon}\sigma\sigma\iota\nu o\varsigma$, whereas in our passage the word used in the received text is $\lambda\acute{\iota}\nu o\nu$, which is never used for fine linen. But this rather strengthens the argument for the received text. For the very reason why βύσσινος is *not* used here is that the *fineness* of the linen is expressed by no fewer than two adjectives —"pure, bright" (καθαρόν λαμπρόν)—so that, were we to translate literally, without inserting the word "fine" at all—thus, "clothed in linen, pure, bright"— the whole idea of the received reading would be perfectly brought out. (4) If still it be asked whether it would not be more natural for copyists to change the harsh word 'stone' in such a passage into the smooth word 'linen,' than the reverse—I answer, 'Yes, if we could suppose this done *intentionally*,' but that we never suppose: we hold it to be a pure mistake—the substitution of one letter for another, giving to one word which in every other letter is identical with another a sense totally different. Bengel's rule—that the rough reading is preferable to the smooth one, *cæteris paribus* —is an excellent one. But to apply it to mere blunders is to abuse it.

A word now on the textual evidence. Of the Apocalypse only five Uncials are known to exist, and only three of the oldest class, ℵ A C, with B_2 (8th century), and P_2 (9th century), and several cursives. Reckoning versions, the evidence is fairly balanced: for the revised text (λίθον), AC, three cursives and the margin of another, the two best copies of the Vulgate, and "some MSS. known to Andreas" (an Apocalyptic commentator of 6th century), but who himself does not adopt it: for the received text (λίνον, variously written λινουν, ληνουν, ληνους, ληνον), ℵB_2P_2, Latin MSS. known to Haymo (9th century), copies of the Vulgate, the Clementine edition of Vulgate, the Syriac and Armenian; and of the fathers Tichonius (4th century), Andreas and Primasius (6th century).

Thus viewed, this repulsive reading has not even such textual evidence as should alone entitle it to a hearing. The more regrettable, therefore, is its presence in the Revised Version.*

---

* Of the critical editors, Lachmann, Tregelles, and Westcott and Hort, read λίθον, while Griesbach and Tischendorf retain λίνον.

FINIS.

www.ingramcontent.com/pod-product-compliance
Lightning Source LLC
Chambersburg PA
CBHW021813230426
43669CB00008B/743